COSMIC RAYS

COSMIC RAYS

THREE LECTURES

being the revision of the
1936 Page-Barbour Lectures of the
University of Virginia and the
1937 John Joly Lectures of
Trinity College
Dublin

by

R. A. MILLIKAN

CAMBRIDGE
AT THE UNIVERSITY PRESS
1939

CAMBRIDGE
UNIVERSITY PRESS

University Printing House, Cambridge CB2 8BS, United Kingdom

Published in the United States of America by Cambridge University Press, New York

Cambridge University Press is part of the University of Cambridge.

It furthers the University's mission by disseminating knowledge in the pursuit of
education, learning and research at the highest international levels of excellence.

www.cambridge.org
Information on this title: www.cambridge.org/9781107689657

© Cambridge University Press 1939

First published 1939
First paperback edition 2014

A catalogue record for this publication is available from the British Library

ISBN 978-1-107-68965-7 Paperback

CONTENTS

PREFACE

These three lectures were prepared as the Page-Barbour lectures of the University of Virginia, and were so delivered in the spring of 1936. The delay in publication has been partly due to multiplicity of other duties, and partly to the discovery, beginning in the summer of 1936, of important new results, the general significance of which it seemed very desirable to include in any publication in book form which was not to be immediately outdated. The lectures were accordingly revised and delivered as the John Joly lectures at Trinity College, Dublin, in the spring of 1937, and have now been brought up to the spring of 1939.

The lecture form has of course precluded any compendious treatment of the immense mass of publications in this field. Those who are interested in a condensed but quite complete and thoroughly objective review of all the work in the field, including a full bibliography of the publications up to the summer of 1932, can find it in the very fine paper by one of the earliest and most fruitful of the cosmic-ray workers, Prof. G. Hoffmann, published in the *Physikalische Zeitschrift*, **33**, 633–63 (1932). Full references up to the end of 1937 (2900 of them) are found in E. Miehlruckel's *Hohenstrahlung*, Steinkopff, Dresden, 1938. Also Thomas H. Johnson's article in *Reviews of Modern Physics*, October 1938, contains an excellent bibliography.

The present lectures of course deal primarily with the results of our own group of workers at the Norman Bridge Laboratory of the California Institute of Technology, but a careful attempt has been made to include at least the most significant contributions of other workers, so that it is hoped that the reader will get a condensed but fair picture of the most important of the present findings in this new and rapidly developing field of cosmic rays.

R. A. M.

2 *April* 1939

The Discovery of Cosmic Rays and Its General Significance

I wish to devote no small part of this first lecture to telling not merely how the discovery of cosmic rays came about, but, more important than that, to attempting to answer the question which every cub reporter asks within two minutes of the beginning of an interview: "What are they good for?" I am going to surprise some of my audience by expressing some sympathy with that inquiry, though no sympathy with the demand to receive an answer that can be made a headline for a tabloid newspaper column, such as "Millikan finds in cosmic rays secret of eternal life", or "Scientist finds in cosmic rays inexhaustible sources of energy". For the question, "What are cosmic rays good for?" is like that other question generally put by a child-like fundamentalist or an equally child-like atheist—two individuals who are actually very much alike in their mental processes—"Do you believe in God?" or again like an equally puerile question recently put to me by one of our most prominent, sincere and unintelligent proponents of the doctrine of Prosperity through Scarcity, namely, "Why are scientists and engineers always conservatives, if not reactionaries?"

All these questions are of the type that require "an education rather than an answer". They spring from the

Santa Claus stage of mental development, and yet they are all capable of an answer if the inquirer is capable of being sufficiently discriminating, and is willing to be sufficiently patient, to start by getting into agreement as to the definition of terms—in a word, to get the general background of the man whom he is questioning.

If one is capable of doing that, and willing to do it too, then I have the greatest sympathy with the question, "What are cosmic rays good for?" and I am going to spend a certain fraction of this hour trying to answer it. For I have no sympathy whatever with the statement often attributed to scientists that it is not necessary or desirable to consider whether or not their work has value for mankind. Whether it has or not, at any rate this much is certain: It must have value from the standpoint of somebody other than the scientist—and we shall call that somebody other than the scientist society as a whole—or else it will not be supported by society. In other words, it is necessary, and I think also proper, to demand of every worker in any field to demonstrate to society that his work has value for it, as well as having the value for him of gratifying his individual pleasure, if he expects society to support him in it.

A simple and extreme illustration will make this altogether clear. Suppose it is my desire to sit and pound a log all day with the butt end of a hatchet. My friends may bring me food for a few days out of a humane interest in me, but if I cannot convince them that I am accomplishing something by my pounding they will eventually see to it that I get my food brought to me in the insane asylum until my brain, deranged from society's point of view, recovers its normality in its estimate of values, or else is mercifully returned to the dust from which it sprang.

The answer to the question, "What are cosmic rays good for?" can best be made by calling on the experience of the past, although in calling upon the past at all I realize that to certain types of beings—I purposely avoid the word "minds"—I am showing myself "a hopeless reactionary". Suppose, then, that we try to see what answer history has given to the altogether similar question, "What is astronomy good for?" For astronomy has never directly filled anybody's stomach or satisfied anybody's material wants of any kind except the astronomer's and his wife's and children's, and they got their spinach and their automobile only because you (society) gave it to them. Did the astronomer make any return to you? In just one way, namely, in adding something to your knowledge of the universe and in revealing to you the *method* by which real knowledge is obtained.

Has that real knowledge then *indirectly* done anything to fill your stomach or buy you an automobile? I dislike to take the time to answer that question because it diverts attention from the point of most supreme importance, but actually it gave you the law of gravitation; it demonstrated for you first in celestial affairs the correctness of the fundamental laws of Galilean and Newtonian mechanics, which nobody ever knew anything about before A.D. 1600. It was two hundred years before you began to gain any material good from those discoveries: but then, about A.D. 1800, two hundred years after Galileo's time, you began to see that you could apply those same principles to the problems of terrestrial mechanics, i.e. to the problems of your own daily life, and as a result in the nineteenth century the industrial revolution came about and this changed the whole face of modern life. Not an automobile or power machine of any kind can be built to-day without applying directly the fundamental scientific principles discovered by Galileo and Newton some three hundred years ago. There have been thousands of additions to their discoveries, it is true, but it is not an exaggeration to say that modern civilization, whatever its value may be, actually rests fundamentally upon the principles brought to light largely through the study of that useless thing, astronomy, by Galileo and Newton and their contemporaries and followers. Had they not made those discoveries, modern industrial civilization could not have developed unless or until others had made these same discoveries. There you have the indirect bread-and-butter value of the study of astronomy.

But in these last paragraphs I have over-emphasized the material. That is after all a minor part of the social values that came from the study of astronomy. The change in man's conception of the universe that came directly from the invention of the telescope and microscope in Holland about 1600 slowly revolutionized man's thinking, revolutionized his theology, his literature, and his politics.

Galileo's discovery with the aid of his telescope having a 1 inch objective (contrast this with the new 200 inch) of a universe of stars beyond, or better in addition to, the visible universe, some of which were seen to be changing, too—Jupiter had definitely revolving satellites brought to light by Galileo's telescope, a new star appeared in the heavens in 1604—these discoveries destroyed in men's minds the mediaeval ideas of the great, fixed, eternal, unchangeable firmament, and in time they released man from the bondage of some of his worst and most unsocial superstitions. Of course mediaeval theology had to change or die. New knowledge showed that some of its fundamental postulates were wrong. The mind of man began to become free, to study for itself, to speak out its convictions, and to act upon them. Authoritarianism imposed from above, both heavenly and earthly, lost its sacredness; the idea of the totalitarian state, dominant since the days of the Pharaohs, began to be questioned; ideas of economic and political freedom began to stir. Says Harold Moulton, Brookings Institution economist, "It appears clearly to have been the influence of the great scientific discoveries of the seventeenth century which in due course provided the philosophical foundations for the system of free enterprise. The key to the great transition from regulated to free enterprise was found in the conception of 'nature's laws' with which the physical scientists were concerned." The imagination of man was stimulated by the newly discovered far reaches of the universe. Literature and poetry felt the effect. A specialist in the literature of that period attributes the difference in the types of poetic imagination revealed in Shakespeare and Milton to the new scientific knowledge about the constitution of the heavens and the earth brought in between their dates by the telescope and the microscope. Milton's angels fell through a new kind of space. *Ideas are after all more potent than machines in determining the direction of human evolution and the fate of empires.*

Now, at no two epochs in the history of the world have

men's minds been so stirred by the influx of new knowledge about the nature of the universe as, first, in that century which stretches forward and backward from 1642, the date of the death of Galileo and the birth of Newton, and, second, in the half century that stretches back from the present, 1938, to the year 1888, the year in which Hertz discovered wireless waves. This just happens to be within a year of the time in which I entered college, so that I can relate what I myself have seen happen since that time.

I propose to take one bit of this new knowledge—that underlying the discovery of the cosmic rays—and show how it has come about, how it has all grown step by step by thousands of little increments out of the body of knowledge and experience accumulated in the centuries preceding, how striking an illustration it furnishes of the effectiveness of the scientific mode of approach to new knowledge and to new avenues of progress, for I suspect that there are no others.

Possibly the largest social value that can come from the study of the cosmic rays will consist in furnishing to our generation another object-lesson of the method by which real progress, scientific or social, is to be obtained. For the new scientific ideas that came into human thinking through the development of the telescope and the microscope in the early seventeenth century undoubtedly had much to do with the shaking off of the shackles of both religious and political authoritarianism and the consequent initiation in the eighteenth century of the world's first great, far-flung experiment in religious, intellectual and political freedom in the founding of the United States, and these ideas have spread rapidly since that time until about the close of the great war. But what is happening now? Never since 1600 has the world seen such a reversion toward authoritarianism, superstition and every irrational and unscientific brand of emotionalism as at the present moment. And wherever emotionalism determines conduct, there you have necessarily the law of the jungle. Let the sentimentalist note well that statement. In it I have given the correct definition of a reactionary, namely, the man who has turned his face back toward the method of the jungle, toward brute government instead of ballot government, toward authoritarianism instead of toward freedom.

On the right and the left are found all the reactionaries, only in the centre is found the real progressive. Why? Because there, as Walter Lippmann has so well said, are found all those who are trying to replace the method of the brute by the method of a being supposed to be endowed with a mind—that is, by analysis, by persuasion, by adjudication, by compromise, by evolution, by peaceful change. On the left, and in somewhat lesser numbers on the right, are found all those who, no matter to what kind of liberalism they may pretend to adhere, actually *support by their influence*, and practice in varying degree, violent world revolution, assassination and intimidation, suppression of freedom of speech, press and action, indoctrination of the public in the interests of the ideas and the individuals at the moment in power, despotism—in a single word, *reaction*. Not for three hundred years have the forces that make for freedom and progress been so sorely pressed the world over as now. The United States is not yet as badly off as are some countries, but it has certainly been influenced by the wave of hysteria and reaction toward political despotism that has swept in succession over Russia, Italy, Germany, and some other countries of Europe. Who would have believed that physical, social and economic perpetual-motion machines and movements could be sold to great masses of people, even in the United States, in this nineteenth century—movements, some of which, such as "Townsend" and "Utopia" plans, "Prosperity through Scarcity", "Abundance through Borrowing", "Distribute the Wealth", etc., violate the laws of simple arithmetic, and the great majority of which have not the endorsement of any of the men who are regarded by their fellows in their own fields as the ablest, most experienced and most informed, most high-minded men to be found, be they physicists, engineers, economists, or statesmen? Obviously, the only scientific way to approach the solution of any problem is to call in the men who have the right to be considered as informed and competent. Who would have believed that movements without any scientific backing or credentials at all could be hawked about the United States in high places and in low, from soap boxes and from the seats of power, and get millions of voters behind them, too?

The cosmic rays seem to be popular with the public just now. Perhaps one of the things they are good for, like the coming of the telescope in 1600, is to help to get a more rational, more scientific, more intelligent mode of approach into public thinking than just now exists in the United States. Perhaps they will help to save freedom and democracy now just as the discoveries of 1600 with the telescope helped mightily to create freedom and democracy then.

Now you know why "scientists and engineers are always reactionaries". It is because they have learned by life-long experience that they cannot improve bridge-building and ignore the fundamental laws of structures which have already been discovered. It is because they have been trained to analyse their problem with the aid of all the knowledge that is available before they court disaster by following merely their feelings, their hopes, or their hunches. It is because they have learned by bitter experience that they cannot forget the law of gravity without doing irre-parable damage. It is because they try to guide their lives by facts and are not very susceptible to "hooey", whether emanating from a soap box, a governor's mansion, or a White House. It is because they have been trained to look behind the surface and to try to get at the fundamental causes of bad conditions and to find how to eliminate them. It is because they have little confidence in the long run in artificial prosperity—in making one think he is all right for a little while by giving him a shot in the arm. And this scientific method can be used, too, in all fields, though unfortunately it is not so used, particularly in politics.

Let the story of the growth in our knowledge of the field of radiation culminating in the discovery of the cosmic rays be an illustration of the method. When I started for college in the year 1887, about fifty years ago, what we knew about radiant energy was largely confined to the visible spectrum represented by the very narrow white area on the logarithmic scale of now thoroughly explored radiation frequencies represented in Fig. 1. How enormous a range of frequencies is there represented will be realized when it is known that each division shown on the right-

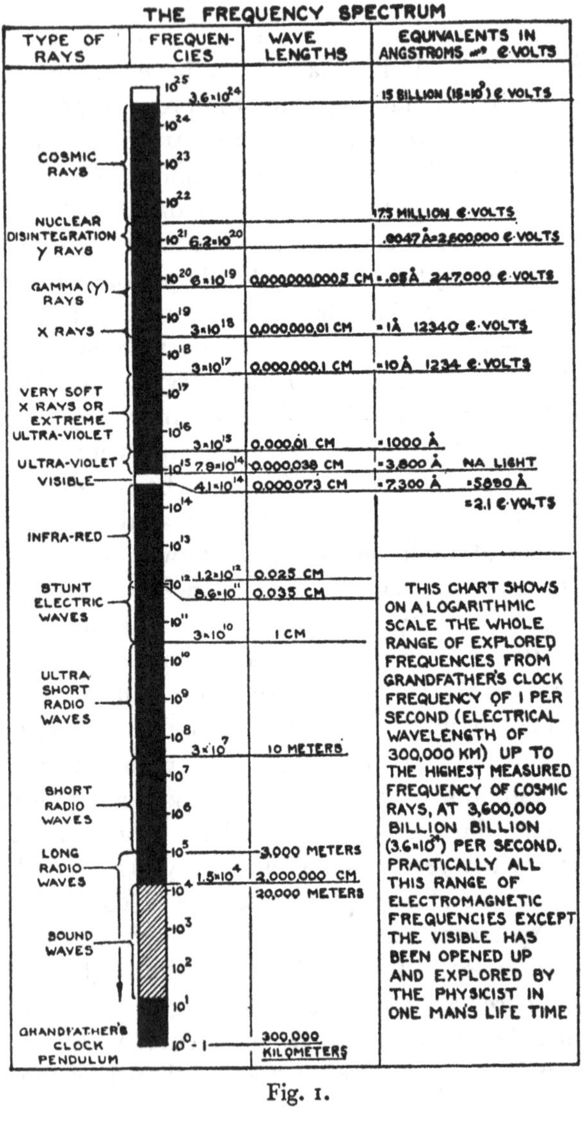

THE FREQUENCY SPECTRUM

TYPE OF RAYS	FREQUEN-CIES	WAVE LENGTHS	EQUIVALENTS IN ANGSTROMS AND E·VOLTS
	10^{25}		
	3.6×10^{24}		15 BILLION (15×10^9) E·VOLTS
	10^{24}		
COSMIC RAYS	10^{23}		
	10^{22}		
NUCLEAR DISINTEGRATION γ RAYS	10^{21} 6.2×10^{20}		75 MILLION E·VOLTS .0047 Å = 2,600,000 E·VOLTS
GAMMA (γ) RAYS	10^{20} 6×10^{19}	0.0000000005 CM = .05 Å 247,000 E·VOLTS	
	10^{19}		
X RAYS	3×10^{18}	0.000,000.01 CM	= 1 Å 12340 E·VOLTS
	10^{18}		
	3×10^{17}	0.000,000.1 CM	= 10 Å 1234 E·VOLTS
VERY SOFT X RAYS OR EXTREME ULTRA-VIOLET	10^{17}		
	10^{16}		
	3×10^{15}	0.000,01 CM	= 1000 Å
ULTRA-VIOLET	10^{15} 7.9×10^{14}	0.000,038 CM	= 3,800 Å NA LIGHT
VISIBLE	4.1×10^{14}	0.000,073 CM	= 7,300 Å = 5890 Å
	10^{14}		= 2.1 E·VOLTS
INFRA-RED	10^{13}		
	10^{12} 1.2×10^{12}	0.025 CM	
STUNT ELECTRIC WAVES	8.6×10^{11}	0.035 CM	THIS CHART SHOWS ON A LOGARITHMIC
	10^{11} 3×10^{10}	1 CM	SCALE THE WHOLE RANGE OF EXPLORED
	10^{10}		FREQUENCIES FROM
ULTRA SHORT RADIO WAVES	10^{9}		GRANDFATHER'S CLOCK FREQUENCY OF 1 PER SECOND (ELECTRICAL
	10^{8} 3×10^{7}	10 METERS	WAVELENGTH OF 300,000 KM) UP TO
	10^{7}		THE HIGHEST MEASURED FREQUENCY OF COSMIC
SHORT RADIO WAVES	10^{6}		RAYS, AT 3,600,000 BILLION BILLION
LONG RADIO WAVES	10^{5}	3,000 METERS	FREQUENCY OF COSMIC (3.6×10^{24}) PER SECOND. PRACTICALLY ALL
	10^{4} 1.5×10^{4}	2,000,000 CM 20,000 METERS	THIS RANGE OF ELECTROMAGNETIC FREQUENCIES EXCEPT
	10^{3}		THE VISIBLE HAS BEEN OPENED UP
SOUND WAVES	10^{2}		AND EXPLORED BY THE PHYSICIST IN
	10^{1}		ONE MAN'S LIFE TIME
GRANDFATHER'S CLOCK PENDULUM	10^{0} = 1	300,000 KILOMETERS	

Fig. I.

hand edge of the heavy vertical band on this chart means a group of frequencies ten times greater than that represented by the division just below it, so that moving up six steps corresponds to multiplying the frequency a million-fold, twelve steps a million-million-fold. The chart furnishes then a very striking illustration of what the scientific mode of approach has made possible in less than one man's lifetime.

Reflect, too, upon the fact that the knowledge merely of *visible* frequencies contained in that little white strip at the middle of the vertical dark band of the chart represents wellnigh a century of analytical and experimental work on light by the ablest minds of the nineteenth century; and that every bit of that knowledge as to the wavelengths, the polarization, the speed of propagation of light, and the interference effects observable in it, had to be used to make the next advance. Indeed, the matchless genius of Maxwell had put all the then known knowledge of light and of electricity into the development of the electromagnetic theory which *predicted* that next advance made by Hertz in 1888. It was then that the whole region of the electromagnetic spectrum, known as wireless waves, was experimentally brought to light and began to be explored. During the following decades these waves were stage by stage proved to be identical with light waves in speed of propagation, polarization and interference effects, and all this knowledge was available for making possible the next stupendous advance in wireless that began in 1915 with the development of "the electron-tube amplifier", which now underlies the whole art of communications, as well as the whole sound picture industry, not to mention a score of minor industries, or the opening up of important new fields of science.

Next, beginning about 1890 the whole field of heat waves lying just below the white strip was explored with the use of all the accumulated knowledge of visible radiations, until now, as the chart indicates, the whole range of electromagnetic frequencies from those of light down to those of any desired wave-length has been added to the physicist's store of definitely accumulated knowledge. It will be noticed that the lowest line on the chart represents a vibration rate of one per second, the frequency of the pendulum of "grandfather's clock".

Then in 1895 through Roentgen's discovery we found rays which differed from all rays thus far known in their extraordinary penetrating power. From 1895 to 1912 we struggled over the nature of these rays, and then again, by using all the interference techniques and all the analysis which our century-long study of light had developed, we definitely proved that X-rays were merely very short wave-length light of wave-length from a thousandth to a hundred thousandth that of ordinary light, as the chart shows.

And then in 1896 came the discovery of radioactive rays stimulated by, and built directly upon, our knowledge of X-rays. And it took seven more years of work, which utilized all of our accumulated knowledge both of light and of cathode rays, the latter of which had been twenty years in the making. Then with the aid of all this accumulated knowledge in the hands of those, and only those, who had had the competence to master it, the riddle was solved and radioactive rays were proved to be separable into three kinds of rays, called α, β, γ rays, the first two of which were deflectable in a magnetic field and therefore consisted necessarily of electrically charged particles, while the last were like light in being undeflected by such a field. These three rays differed also in penetrating power, the α-rays (charged helium atoms) being about 1/100th as penetrating as the β-rays (streams of high-speed electrons), while the latter were 1/100th as penetrating as the γ-rays. The last were guessed to be, though not proved to be until after 1912, simply shorter wave-length light than X-rays; and so another region of ultraviolet radiations was added to our frequency chart. All this new knowledge of frequencies in the ultraviolet had been built directly upon the knowledge of interference, both as to theory and experiment, developed in the first half of the nineteenth century. In other words, we were simply building upon and adding to, bit by bit, the accumulated knowledge of the past.

The next big step that was taken started with the unexpected discovery that light waves have bullet properties. This looked like a contradiction in terms, and we at first thought either that the old knowledge or the new experiments must be wrong. We called on all our old techniques to prove that either our bullet effects in electromagnetic wave phenomena or else our interference effects were fundamentally incorrect. But real, experimental knowledge can never be unhorsed, and after tilting at it with our Don Quixote lances for almost two decades we accepted the evidence of facts, adjusted our minds to this kind of dualism in the nature of light, and went on with our explorations.

Our modern photon theory of light and of all electro-magnetic waves tells us that each element of light is a localized bundle of electromagnetic energy which under all circumstances travels through space as a bunch of con-centrated energy at the invariable speed of 186,000 miles a second, the amount of the energy carried by each bullet, which we shall henceforth call a *photon*, being proportional to its frequency as measured by the interference effects shown in an ordinary prism or grating spectroscope. These light bullets, or light darts, for some reason which we are not yet able to state in physical terms, always distribute themselves in the interference pattern given by the classical wave theory. This, at least, is the experimental situation to which we must henceforth adjust our thinking.

For our present purposes this is a very convenient dis-covery, for it means that if we have measured the frequency ν of a light ray by the usual methods we know at once its energy, E, in ergs from the equation $E = h\nu$, where h is a universal constant of known value called "Planck's h". Inversely, if we have any means of measuring energy we can obtain at once the frequency from the same equation. This brings me to the discovery and the measurement of the frequency of cosmic rays, for we have not measured their frequencies directly, but we have measured their energies; and that is how the upper end of this amazing chart of our expanding knowledge has been filled in at the top in the manner shown.

Prior to the discovery of the cosmic rays no one could have been easily persuaded that any frequency very much higher than those of the γ-rays of radium (see chart) could exist. The reasoning would have been as follows: We have just learned that we can at once write down the energy of a ray of sodium light of wave-length about 0·0006 mm, or 6000 ångströms. It comes out close to 2 electron-volts, the electron-volt being the energy that a single electron acquires in falling through a difference of potential of 1 volt. In the same units the energy of an X-ray photon having a wave-length of 1 ångström, such as a small laboratory induction coil produces, is about 12,000 e-volts, for frequency is of course inversely proportional to wave-length; and so if the wave-length that we are considering is 1/6000th that of sodium light, the frequency will be 6000 times greater and hence the energy will be 2 × 6000 = 12,000 e-volts, as the chart shows. Now, the modern theory of radiation indicates that whenever an atom of sodium emits a 2 e-volt photon it is because some electron in one of the outer electron orbits or shells of the sodium atom has fallen from some more remote orbit, or position, in toward the nucleus, and in so doing has lost just the 2 e-volts of atomic energy that shot out in the form of the 2 volt photon. The electrons in all the electronic shells or levels of the various atoms are bound to their nuclei with a definite amount of binding energy, which is now pretty well known, and the most powerfully bound electron in any known atom is one of the two inmost electrons of the so-called K shell of the heaviest of all atoms, the atom of uranium. It takes a blow of just about 100,000 e-volts of energy to knock one of these K electrons out of its home position within this atom, and when this, or some other electron, jumps "back home" a 100,000 e-volt photon will be emitted.

The so-called "characteristic" X-rays that have their

origin in the tungsten target of an X-ray tube are supposed to be generated by just this process. Cathode rays are in this case shot from the hot filament into the tungsten target with about 70,000 e-volts of energy. Some electron in the K shell of the tungsten atom is thus knocked out, and when this or some other electron falls back into the vacant hole a 70,000 e-volt photon is emitted by the target, and with its great penetrating power goes through your jaw and reveals therein the abscessed tooth.

But the γ-rays of thorium spontaneously emitted by the nucleus of the atom of thorium just after an electron has shot out, not now from one of the surrounding electronic shells but from the nucleus itself, and left some kind of a hole in that nucleus, is found to have an energy of as high as 2·6 million e-volts. It was natural to assume, therefore, that a few million e-volts of energy was the highest to be expected from *nuclear* changes of any kind, so that a decade or two ago no one expected that higher energy photons than those having an energy of a few million e-volts would ever be found. Certainly none such had ever appeared on earth so far as anybody knew. The highest energy with which α- and β-rays shot out of the nuclei of radioactive atoms had been measured at about 8 or 10 million e-volts, but this was the atomic bullet, whether photon or electron, of largest energy that could exist, *so we thought*. But here again we were wrong.

The radioactive substances uranium and thorium, which emit rays of this sort, were known to be widely scattered in minute concentrations, as Joly and Poole had first shown, throughout the earth's crust, and while very penetrating γ-rays were found as early as 1902 and 1903 entering electroscopes placed almost anywhere on the earth's surface, they were universally interpreted as having their origin in this wide distribution of uranium and thorium throughout the earth's crust. Then in 1909 Göckel in Switzerland took up an electroscope in a balloon two or three times and found it always discharging faster at a height of 4 kilometres than at the earth's surface. This was completely inconsistent with the hypothesis of an origin in the earth's crust. The next year the Austrian, Hess, fittingly awarded the Nobel prize in 1936 for his early contributions to the cosmic-ray field, repeated, checked, and extended Göckel's balloon observations and cautiously advanced the hypothesis that these high-altitude rays originated outside the earth, though he recognized that their origin in the atmosphere had not been excluded by his observations. He also observed that the discharging effects seemed to be independent of the position of the sun. The German, Kolhörster, in the next two years extended these balloon observations to an altitude of some 29,000 feet and found the discharging effects at those altitudes some eight or ten times greater than those found on earth, and again with due caution interpreted his results in terms of a very penetrating radiation coming in from outside—very penetrating because, if it did come in from outside, then in order to get through the atmosphere and make itself felt down here close to the earth's surface it had to be from five to ten times as penetrating as the most penetrating radioactive rays. The uncertainty about these conclusions so far lay in the fact that all that had been proved was that the rays discharged electroscopes much faster at high altitudes than at low, a result equally well interpretable in terms of a suitable distribution throughout the atmosphere of rays of an ordinary radioactive penetrating power, or in terms of rays of much greater penetrating power originating outside the atmosphere.

The next significant experiments were very high balloon flights, up to 55,000 feet, made in 1922 by Bowen and Millikan, with self-registering electroscopes launched from San Antonio, Texas. These were the first stratosphere flights made with electroscopes. They showed that the rate of increase in the ionization did not continue constant up to the top of the atmosphere and therefore seemed to us to point toward some kind of an ordinary radioactivity suitably distributed through the upper atmosphere and to point away from an extra-terrestrial origin of a very much more penetrating radiation than could be accounted for by radioactive transformation. But by the year 1925, both Kolhörster and Millikan and Cameron had made direct measurements of penetrating power, the former by measuring the intensity of the ionization on top of an Alpine glacier and then in a crack some distance beneath the surface; Millikan and Cameron by sinking electroscopes stage by stage to depths of 70 feet beneath the surface of the waters of Muir Lake (California), which they had proved to be radium free, and noting a continuous diminution in the rate of discharge down to that depth. These experiments proved unambiguously the existence of rays penetrating enough to go through a thickness of water some three times the equivalent of the earth's atmosphere, and if penetrating power were a measure of energy, then of much greater energy, or frequency, than any rays found on earth. Millikan and Cameron also proved to their own satisfaction that the rays could not originate in the atmosphere by proving that an equivalent thickness of the earth's atmosphere weakened these rays precisely as did the corresponding thickness of radioactive-free water. Neither Kolhörster's evidence, however, nor Millikan and Cameron's was sufficient to convince all physicists, one of the best known of whom insisted at least up to 1927 that the upper layers of the atmosphere might be the place of origin of these penetrating rays. This point of view is now, however, completely abandoned, for both of the following arguments seem unanswerable. First, the remoter reaches of the sun's atmosphere ought to be such a source of rays if the earth's outer atmosphere behaves in that way, but this suggestion is completely negatived by the practical equality of the rays by day and night, i.e. as the earth turns its face toward or away from this supposed source. Secondly, the effect, to be later detailed, of the earth's magnetic field is by common consent completely irreconcilable with the view of the upper atmospheric origin of the rays. One view was wrong, the other right. There was no uncertainty about it. The rays were first called cosmic rays in the report of the 1925 Millikan and Cameron experiments in snow-fed lakes.

The first definite proof that they come from beyond the Milky Way was brought forward in 1926 when experiments were made in South America where the Milky Way is completely out of sight for hours at a time, and the proof there found that the intensity of the rays is just the same when the Milky Way is out of sight as when it is in full view. The same is true to a first order of accuracy about the position of the sun, so that it would seem that neither the sun nor any of the stars in our galaxy can possibly be a significant primary source of origin of the cosmic rays. But what, then, lies beyond the Milky Way that can possibly act as a source of these extraordinary influences?

Fifteen years ago no one could answer that question. No one knew that there was any such place as "beyond the Milky Way", but within that time we have learned to measure quite accurately the distances of very remote stars and within the last ten years have learned much about what lies beyond the Milky Way.

Nevertheless, we do not yet know what kind of celestial bandits, or, if one prefers so to speak, what kind of impersonal events, out there beyond the Milky Way are responsible for this continuous shooting up of our dwelling-place, the earth. We have learned a great deal about the kind of bullets they use and how these affect different kinds of substances, and the next two lectures will deal with this knowledge, with how we have obtained it, and to what new ideas it gives rise; but it is very important for understanding the social significance of science—which is really what this lecture has been mostly about—to realize that there are many questions even in physics of which we do not yet know the answer, and the mechanism of origin of the cosmic rays is one of them.

If we represent the field of physics by a sphere, then there is at the centre of this sphere a core of definitely established knowledge. By this I mean simply knowledge upon which 95% of competent and informed physicists will be in agreement. I leave 5% to cover the case of the man found in all fields who is so constituted that he never agrees with anything or anybody, for this case is outside the domain of physics and has entered the field of pathology. Surrounding this core of practically established knowledge there is a zone of uncertainty wherein practically all the controversies among physicists occur. Indeed, the progress of physics consists in nothing other than the gradual building up of this inner core through discussion, analysis, and research at the expense of the nebulous shell of the uncertain and the controversial surrounding it. And then again outside this nebulous or twilight zone lies the dark and vast unknown.

Now, I am not presumptuous enough to assume that it is only the physicist who in the last few centuries of intensive intellectual activity, since the discovery of the scientific method, has established his core of fundamental knowledge. In some fields the core will of course be larger than in others, but it is incredible to me that in economics, in government, in finance, indeed in every field of human effort, the experience and the intensive study by men quite as intelligent as the physicist have not brought into existence a core of established knowledge, the fundamentals upon which future progress must be based, and the violation of which can only lead to disaster. I am only a physicist, but perhaps I might be bold enough to illustrate by mentioning one or two of these social *fundamentals*. Socrates, the least dogmatic of all philosophers, stated one of them and drank the hemlock rather than violate it. He knew, despite his continual assertion of lack of knowledge, that organized society could not endure at all unless the individual citizen had respect for, and was willing to live in conformity with, the rules which a free community had adopted as those under which it would conduct its organized life. For him it was not a question of the wisdom or unwisdom of The Laws. It was only a question of living in accordance with the agreed upon rules or of changing them *by the established and recognized procedures*. And he chose to die unjustly rather than to permit the influence of his example to weaken the sanctity of The Laws. The *Phaedrus* of Plato should be re-read to-day by every thoughtful person in the United States, for in it Socrates gave, twenty-three hundred years ago, the final answer to the Supreme Court issue which has recently been raised in the United States.

Another social fundamental was stated by Cicero in these words: "He who violates his oath profanes the divinity of faith itself." When modern governments consider their treaties as scraps of paper, and heads of governments lightly violate their public promises and their oaths of office, they strike a blow at the very foundations of our whole social structure, for public perfidy creates an atmosphere which poisons private honour. Nothing but disaster can follow the violation of such fundamentals.

A similar fundamental in the field of economics is that it is impossible to distribute more of goods and services than are produced, a corollary to which is that anything that reduces the total amount of goods and services produced necessarily lowers the standard of living of a people. If the average voter were fully aware of that fundamental and its implications, how the crowd around the average soap-box orator would disperse, and how the radios would be turned off when the usual demagogic speech comes over the air!

Of course it is true that not all the questions that come before the average man for decision are covered by these fundamentals of established knowledge, but in my own case I have been amazed to find how many of them are. The majority of the questions that come from the public to us in the physical laboratory have to do with perpetual-motion machines or similar matters, to which a definite and final answer is already available, and I suspect we are cursed with quite as many social perpetual-motion schemes as physical.

It is indeed a great social need to gain more knowledge, but it is if anything an even greater one, especially in a democracy, to *find a way by which the knowledge that has already been acquired is made available to the average citizen when he has need of it.* This does not mean spreading it abroad so widely that it has become the property of *every* citizen. That is a desirable but an obviously impossible ideal. Efforts toward it are commendable, but the practical problem is far simpler and not unattainable either, if proper attention is given to it in our public schools. For it is simply a matter of education. The fundamentally educated man is not the man who knows everything himself, but simply the man who realizes his own ignorance and limitations and has acquired the habit and the technique of going to the proper sources for the knowledge which he needs for his decisions. This is not an impossible ideal to put even before the average citizen. If the question is one of physics, it is not an impossible or even a difficult thing to find out whether nine-tenths of the most competent physicists are agreed upon the answer. If they are, there is not a chance in a thousand that they are wrong. Even if they are not in agreement, the best judgment obtainable from physicists is surely the best that is to be had on a problem in their field. And of course the principle is the same regarding medicine, or law, or economics, or government, or any field whatever of human thought and life. Crudely stated the principle is, if you have the tooth-ache go to the dentist, not to the blacksmith. In many matters we already follow it, but much too often we fail to do so, especially in politics, where, in Mr Mencken's classic phrase, we have sometimes "turned over the ship to the ship's barbers".

The supreme social value of science is in *the demonstration which it furnishes of the difference between the right and the wrong answer*, in the proof that it supplies of the tragic fallacy of the dictum that one opinion is as good as another, in destroying the terrible superstition that any opinion that is based only on hunch or emotion or preconception instead of on knowledge has any right to be entertained at all, in the dissemination of the conviction that science has found a method by which truth can be discovered and error in time destroyed, a method by which it is possible to attain in time to the ideal of Socrates, the living of "a reasonable life".

LECTURE TWO

Superpower Particles

In our early cosmic-ray studies we found by direct measurement that these rays had great penetrating power, and we tried to compute the energy of the individual cosmic-ray bullets from their observed penetration, but we had no way of knowing with certainty how penetrating power and bullet energy were related. In following out the best theories that we had to tie these two properties together, we had experimental checks upon our theories up to the value of the particle energies which are found in the radioactive processes taking place on earth, i.e. up to a few million e-volts. But it is always very unsafe to draw conclusions in fields that are beyond the range of experimental test, and the early errors which we made, along with Jeans, Regener, and others, in trying to interpret the origin of the cosmic rays were due to the use of certain formulae worked out and experimentally tested for low-energy rays, but which were later proved to be completely invalid for cosmic rays.

The crying need was for *direct measurement* of the energies of cosmic-ray bullets. Whether these were, in their origin, photons or electrons or both was immaterial for the purpose immediately in hand, namely, that of finding by direct measurement the energies of the individual rays; for, as indicated in the preceding lecture, the energies of photons and electrons are interconvertible. Indeed, in general a photon only makes its presence known by transforming its energy, in whole or in part, into that of an electron, the track of which we have means of rendering visible or otherwise detecting.

There are two processes by which this transformation of energy between photons and electrons can take place. In the first, which we call the photoelectric process, the photon falls upon the atoms of ordinary matter and transfers its energy *in toto* to one of the electrons contained in the matter traversed, so that this electron flies out of the atom endowed with the full energy of the incident photon, save for the small amount of work which may have been necessary to detach the electron from its moorings within the atom.

The second process is that of so-called Compton scattering. In this process the photon simply makes impact with an electron in precisely the way in which one elastic ball makes impact with another of different size. Here the energy distributes itself between the two balls after the ordinary laws of elastic encounter, but when the energy of the impinging photon is very high—more than a few million e-volts (and it is only with such energies that we are concerned in the study of cosmic rays)—much more than half of the incident photon energy, on the average, is transferred to the struck electron, so that we can get a pretty good idea of the energy of the incident photon by measuring the energy imparted by it to the electron, provided we do it enough times to get a respectable mean. In brief, then, when γ-rays fall upon an electroscope they make their presence known only by transferring their energy, by one or the other of these two processes, to electrons, which then shoot through the electroscope or other detecting instrument, ionizing as they go. This process is rendered very beautifully visible in the so-called C. T. R. Wilson cloud-chamber technique, as shown in the following figures.

The first of these figures (Fig. 2) shows the tracks of a few electrons that have been jerked out of atoms of air by the passage of a narrow pencil of X-rays through the cloud chamber when no magnetic field was present. The chamber was full of air saturated with water vapour. The sudden movement of a piston connected with the chamber produces a slight expansion of the volume of the chamber. The cooling of the gas caused by this sudden expansion produces a super-saturated condition of the water vapour so that it tends at the instant of expansion to condense. If, then, an electron, released by the impact of a photon, chances to be passing through this chamber at that instant, each of the ions which it forms by the knocking off of electrons from the atoms through which it passes acts as a centre of condensation for a minute water droplet, so that the path of the electron is traced by a line of these droplets. If the energy of the electrons is sufficiently high, in the absence of a magnetic field each one moves in a straight line from the point of origin. In Fig. 2 the electrons have only 30,000 e-volts of energy and hence they are deflected somewhat from their straight path by the forces exerted upon them by the atoms through which they pass. With high electronic energies such deflection is not noticeable in air. To measure high electronic energies accurately it is only necessary that the electron in question be made to travel through a magnetic field, the lines of force of which are at right angles to its path. From the curvature of this path produced by the magnetic field and the known charge on the electron the energy of its motion can be at once deduced. The principle is the same as that involved in the operation of the ordinary electric motor, each conductor in the armature of which experiences a force at right angles to the current flowing in it and also at right angles to the lines of force of the magnetic field.

It was in 1929, after the penetrating power of the cosmic rays had been quite well worked out by observing the rate of decrease of the ionization in an electroscope as the latter was lowered down to depths of as much as 300 feet in the waters of snow-fed mountain lakes,* that the plan matured of making energy measurements of the cosmic-ray particles by the use of a vertical cloud chamber in a very powerful magnetic field. Dr Carl D. Anderson was called in to assist in this undertaking.

The apparatus designed by Anderson and Millikan is shown in Fig. 3. Since it was desired to measure the energy of electrons coming in vertically, the piston was here arranged to move horizontally as shown in Fig. 4 instead of vertically as in preceding experiments with cloud chambers. The shape of the magnetic field and the size of the cloud chamber were so chosen that we felt sure we could measure

---

* Regener later, by similar measurements made in Lake Constance, found minute traces of these rays down to depths of 220 metres. With counter measurements traces at even greater depths have been reported by Barnothy and Ferro, *Phys. Rev.* **53**, 848 (1938), and Volney C. Wilson, *Phys. Rev.* **53**, 337 (1938), both of whom seem to get with counters at more than a thousand metres of water cosmic-ray effects less than 1/10,000th as great as the counts at the surface.

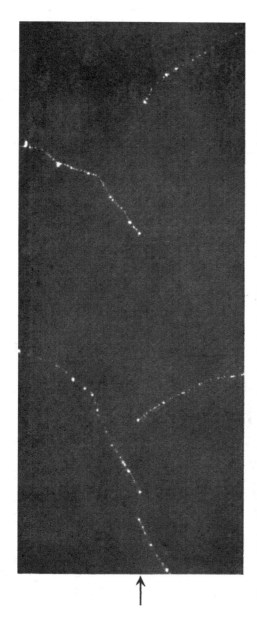

Fig. 2. These beta-ray tracks were taken by C. T. R. Wilson of Cambridge, England. The electrons are photoelectrically released and hence receive the full energy of the incident photons, here fully 30,000 electron-volts. The X-ray beam was here a very narrow one, passing from left to right in a line at the middle of the figure (see the arrow).

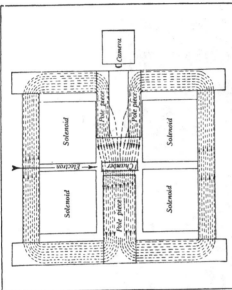

Fig. 3. Extending the measurement of electrically charged "bullets" from 10,000,000 to 10,000,000,000 electron-volts. This is the exceedingly powerful electromagnet built (1930–31) by Carl D. Anderson, with which the energies of the cosmic rays were first directly measured. Electrons plunge from above through the cloud "chamber" (see diagram at right), trigger the "camera", and take their own photographs. As the apparatus is used, the tracks are deflected to the left (in the left-hand image of the following figures) if the electrons are negative, to the right if positive; and by the amount of this curvature we determine their energy. The magnetic field passing through the chamber is produced by a current of 2000 amperes passing through the solenoid. (See diagram at right.) It is essentially uniform and has the huge intensity of 24,000 units (gausses) throughout the whole volume of the cylindrical chamber, which is 17 centimetres in diameter (from top to bottom). It is a simple application of Oersted's discovery of the effect of an electric current in creating a magnetic field around it.

particle energies up to not less than a billion e-volts, this being the order of magnitude of the energies that we expected from our measurements on cosmic-ray penetrating power in water. We actually found it possible with our apparatus to measure electron energies with some precision up to 6 billion e-volts, and to make certain that some few electrons came through the cloud chamber with energies as high as 10 billion e-volts. This was approximately a thousand

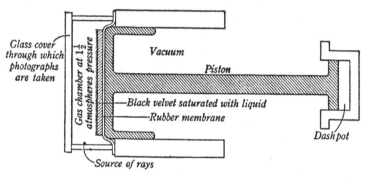

Fig. 4. The Wilson cloud chamber as used at Pasadena for photographing cosmic-ray tracks. For this purpose the chamber is placed vertically in a very powerful horizontal magnetic field.

times higher than the highest particle energies theretofore measured. The coils of the electromagnet were hollow copper tubes through which cooling water continuously flowed. When these coils carried 2000 amperes of current the field was 24,000 gausses over the whole of the "cloud chamber", which was a glass cylinder having a vertical circular area 17 cm in diameter. Its length in the direction of the horizontal lines of magnetic force that passed through it was but 4 cm. The magnet was made particularly light

and well adapted to transportation so as not to be limited
to operation in one place, and actually in the summer of
1935 it was taken, with a special power plant to supply it
with current, to the top of Pike's Peak, and 10,000 photographs were taken there of the energies and other properties of the cosmic rays existing at this altitude. Also,
at least 100,000 such photographs have been taken at
Pasadena and, through the kind assistance of the U.S.
Navy, some 7000 at the submarine charging station near
Colon in the Panama Canal Zone. The main results obtained from a study of these photographs are listed here-
with, combined with those found by other observers, whose
work will be indicated in the references.

1. *The discovery of the positive electron, or positron.*

The discovery of most far-reaching significance resulting
from such cloud-chamber measurements of cosmic rays is
that of the existence of the positive electron, the identical
twin, save for the sign of its charge, of the negative electron
which exists in such numbers in the successive shells
surrounding the nucleus of the atom. The existence of
this new positively charged particle in the world of physics
is brought to light clearly in the following photographs
(Figs. 5–8). Its discovery was made by Carl Anderson in
August 1932 while he was taking photographs with the
foregoing apparatus. It was published in September 1932.
No one had before this even imagined the existence of the
positive twin of the negative electron. Indeed, the proton,
or nucleus of the hydrogen atom, of mass 2000 times that
of the free negative electron, had up to this time been
assumed to be the sole way in which positive electricity

Fig. 5. An electron pair formed by the impact of a cosmic-ray photon on a nucleus of an atom of the gas in the cloud chamber. The left-hand image is the direct one; the right one is formed by reflection from a suitably placed mirror. Stereoscopic examination shows that this positive-negative pair was formed not in the wall, but in the gas itself, and the absence of a track above the point of origin then shows that a photon, not an electron, produced the pair. The track that bends to the left in the direct, or left-hand image, is the negative electron, the other the positive. From the curvatures of the two tracks the energy of the incident photon is found to be about half a billion e-volts.

facing p. 30

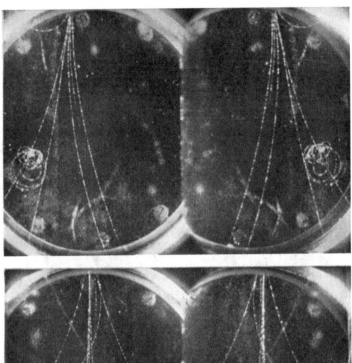

Fig. 6. In the upper picture a 500,000,000-electron-volt cosmic ray hits a nucleus of an atom in the brass piece above the chamber, and three positive and three negative electrons result (positives bend to the right in the direct left-hand image). The circular tracks are due to about four-million-volt photons, which either result from or accompany the foregoing nuclear encounter and an instant later pick off extranuclear electrons from the atoms of the gas in the chamber. In the lower picture a 2,000,000,000-electron-volt ray makes another nuclear hit and produces two positives and six negatives. These pictures were taken with the apparatus shown opposite p. 29.

could exist at all, so that this discovery forced us to change radically our conception of the ultimate constitution of matter. What the cloud-chamber experiments showed was, first, that positive electron tracks frequently appeared as a result of the encounters of cosmic rays with the nuclei of atoms, and secondly, about six months later, that a positive-negative electron pair appears when a photon impinges with sufficient energy upon the nucleus of an atom, particularly a heavy atom. Fig. 5 shows very clearly such a pair produced by the impact of a half-billion volt cosmic-ray photon on the nucleus of one of the atoms found in ordinary air. Also, the formation of other such pairs is shown in Figs. 6–8.

This pair-formation theory was first suggested in the Blackett-Occhialini paper of March 1933* to account for the observed appearance of positive electrons in cosmic-ray photographs. The appearance of such pairs as a result of photon encounters with nuclei was first proved *directly* and unambiguously by Anderson when he was taking cloud-chamber photographs with the aid of γ-ray photons from ThC″ (see Fig. 8). In this case the experimental evidence was excellent that the absorption was here of the photoelectric sort in the sense that *the whole* of the energy of the 2·6 million e-volt photon appears in the electron pair produced. This evidence was found in the fact that the negative electrons that are jerked out of the extra-nuclear shells by the Compton process, to which some of them are bound with very little energy, approach,

* See *Electrons (+ and −), Protons, Photons, Neutrons and Cosmic Rays*, Univ. of Chicago Press, pp. 335–42, for the history of this discovery, with references to the significant papers.

as they should, as a large number of observations on the
energy of different electrons is taken, a value equal to the
full value of the incident photon energy, namely, 2·6
million e-volts, whereas the maximum value of the energy
of the positrons, or the combined energy of a pair when
both are measured, is but 1·6 million e-volts, i.e. about a
million e-volts less, this million e-volts being the equivalent
energy of the electron pair produced by the impact on
the nucleus of the 2·6 million volt energy of the incident
photon. Similarly, Fig. 10 shows that when the energies
of the electrons ejected by the 13 million volt γ-rays
discovered by Lauritsen are measured in this same way,
the maximum energies of the positives is a million volts
less than that of the negatives, this million e-volts having
been expended in the creation of the electron pair.* This
process thus brought to light† of the complete transformation of the energy of a photon into a pair of electrons has
become known as a *materialization* process, electromagnetic
radiant energy being in it wholly transformed into electrons
with definite kinetic energies (material entities).

It is also of much interest and significance that the
evidence is equally good that the full transformation of
energy in the opposite sense may under suitable conditions
take place. Thus, whenever a positive electron has lost
its high kinetic energy by the production of ions along
its path, the attraction between it and one of the negative

* Crane and Lauritsen, *Report on International Conference on Physics,*
I, Nuclear Physics, pp. 130–40 (London (1934)).

† See Anderson and Neddermeyer, *Phys. Rev.* **43,** 1034 (1933);
also Curie and Joliot, *C.R.* **196,** 1105 (1933); also Chadwick, Blackett
and Occhialini, *Proc. Roy. Soc.* **144,** 239 (1934).

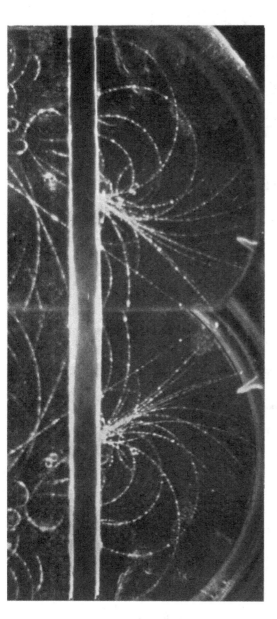

Fig. 7. A 1·5-billion-electron-volt cosmic ray. This picture, taken with the powerful magnet opposite p. 29, shows the results of the collision of a cosmic-ray photon with the nucleus of an atom of lead in the middle of the lead bar 1 centimetre thick. At least five positive electrons (the tracks turned to the right in the left-hand image) and twelve negative electrons (the tracks bent to the left) result from the collision. Positive electrons appear only when an atomic nucleus is hit. The energy of this cosmic-ray photon, as measured by the sum of the energies of all these electrons, to which it transfers both its energy and its direction, is here about 1,500,000,000 electron-volts. Such cosmic-ray bullets shoot through the head of each one of us many times each minute. Indeed, a cosmic-ray ion counter responds about 1·5 times per sec for each sq cm of its area. This means that about 350 shots of one sort or another pass per minute through a circular area 7 inches in diameter—a normal human head.

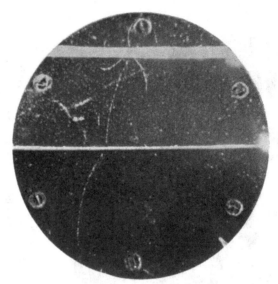

Fig. 8. A positron ejected from a lead plate by gamma rays and passing through a 0·5 mm aluminium plate. The energy above the aluminium is 820,000 volts, below 520,000.

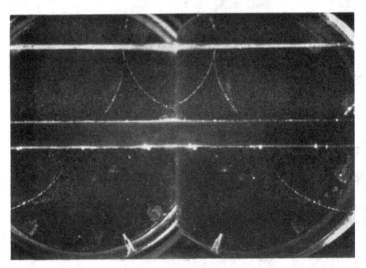

Fig. 9. A striking photograph taken by Anderson and Neddermeyer of a positron-negatron pair of nearly equal energy—about half a million volts each—ejected from a thin lead plate by the impact of a gamma-ray photon of energy 2·6 mev from ThC″. The central centimetre bar is of carbon. The track in the lower half of the picture has its origin in this bar. It is in no way associated with the pair generated above in the lead.

electrons near which it then finds itself, for such electrons are exceedingly abundant in space filled with matter of any kind, causes the two to rush together and mutually

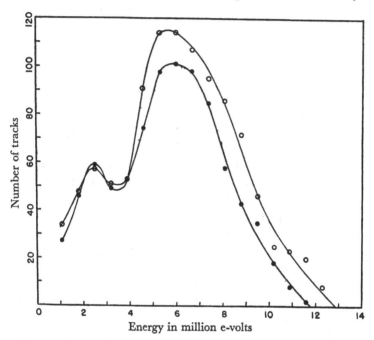

Fig. 10. Energy spectra of the negative and positive electrons ejected from a thick lead plate by the gamma radiation from lithium bombarded with protons. Circles indicate negative electrons and dots indicate positive electrons. Each point represents the number of electron tracks in a 1·4 million electron-volt energy interval. Measurements taken by Lauritsen and his associates.

annihilate each other. The "rest energy" of these two electrons, by virtue of their mass, is the equivalent of about a million e-volts. It is computed from the so-called

Einstein equation, $E = mc^2$, where m is the mass in grams that is to disappear as rest mass (in this case the mass of the two electrons, roughly 1/1000th that of the hydrogen atom), c is the velocity of light in cm per sec, and E the radiant energy in ergs that is to appear. But this radiant energy cannot possibly appear as one single photon of energy 1 million e-volts without violating the third law of Newton that "action and reaction must be equal and opposite"; and indeed the *experimental* evidence* is convincing that two photons each of energy of half a million e-volts do in fact shoot out in opposite directions from the point of annihilation of the positive and negative electrons. So that in this process mass is wholly transformed into radiant (electromagnetic) energy and the applicability of the law of conservation of momentum, Newton's third law, to this process is established.

Some of our cloud-chamber photographs (see Figs. 11 and 12) appear to show, too, that the impact of an electron, as well as that of a photon, can give rise to an electron pair, or pairs. It is difficult, however, if not impossible, directly to determine whether this takes place as a single act or a multiple act, as follows:

When, in an ordinary X-ray tube, the electrons constituting the cathode rays are suddenly stopped by the target, we know that "general" or "scattered" X-ray photons are produced. These are usually called "impulse-radiation photons" or "bremsstrahlung" photons. When

---

* This was proved simultaneously by Thibaud, *C.R.* **197**, 1629 (1933), and by Joliot, *ibid.* p. 1622. See *Electrons* (+ *and* −), pp. 354 et seq., for historical review, including the quite accurate measurements of Crane and Lauritsen.

Fig. 11. Pike's Peak, 7900 gausses. An example of an electron losing a large amount of energy in a lead plate. In this case a positron of ~ 480 mev* energy strikes a 0·35 cm lead plate. Below the plate three electrons appear of energies, respectively: positron, 45 mev; negatron 45 mev; and positron 31 mev. One of the tracks below the plate presumably represents the incident positron after passage through the plate, and the other two tracks a pair generated by the absorption of a photon produced in the plate. The energy lost in the plate by the incident positron is at least 435 mev, which corresponds to a specific energy loss of ~ 1150 mev/cm of lead since the distance traversed is 0·38 cm. The short heavy track may represent an alpha-particle arising from contamination in the plate.

* Million electron-volts.

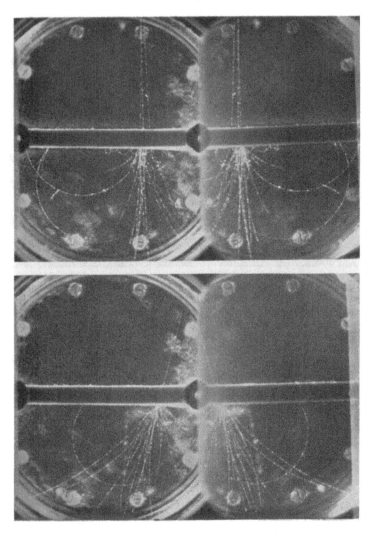

Fig. 12. These figures show an electron or electrons, rather than photons, plunging into a bar of platinum and a number of electron-pairs issuing from the lower side of the bar.

a cosmic-ray electron plunges into a lead plate the first act may be simply the transformation by this process of a considerable part of its energy into an impulse-radiation photon, which, then, by a subsequent encounter with another nucleus transforms its energy into an electron pair. These pairs then in turn produce impulse photons of lower energy, etc., etc.

2. *Cosmic-ray showers.*

The foregoing is the basis of the so-called Oppenheimer-Bethe-Heitler theory of the production of cosmic-ray "showers", a number of which are shown in Figs. 5–14. If the incident energy is in the form of a photon, as soon as it makes a nuclear encounter, its energy, according to this theory, is transformed into that of an electron pair (Figs. 5–9), which may divide the energy between the positive and the negative in any way whatever, the equality in the two energies having the largest probability, but with the possibility that one or the other may get nearly all of it. (In Fig. 5 the positive actually takes some 99 % of the energy.) These two electrons in turn produce new impulse-radiation photons of lower energy, and these produce new pairs of still lower energy, until, by a continuous repetition of this process, the energy becomes divided between a very large number of β-rays, which finally fritter it away through the ionization produced in their ion tracks.

If it is an electron instead of a photon (Figs. 11 and 12) that first enters the absorbing medium, the process is just the same as above after the first event, which consists in the transformation of the energy into that of an impulse-radiation photon, a process which, as both theory and experiment show, becomes more and more efficient the higher the energy of the incident electron. In order that these two events, namely, the transformation of the energy of the high-speed electron into an impulse photon and the subsequent formation of the pairs, may take place close to each other in the absorbing medium, for example, lead, it is obviously necessary that the absorption coefficient of both electrons and photons be very high. Some of the photographs herein shown (Figs. 5–13) are at least consistent with such a theory of shower production. Fig. 13, for example, according to this theory was caught at the very lowest end of this process where the initial energy had become degraded into an immense number of photons and electrons of very low energy, those in the photograph being from half a million to 15 million e-volts, some eighty of which were absorbed *in the gas itself*. In view of the small density of the gas as an absorbing medium, it is to be assumed that here an enormously larger number were absorbed in the surrounding solid bodies, some of which were lead.

3. *"Space-associated" and "time-associated" tracks.*

When one takes photographs with the Pasadena apparatus by the random method, i.e. by merely making exposures until one finds a cosmic-ray track or tracks upon the film, he must make on the average about twenty exposures to obtain a cosmic-ray track, and of these some 15 % are showers.

In the foregoing statement one is defining a shower as two or more tracks which from their sharpness can be

Fig. 13. The cloud chamber happened in this photograph to be at the bottom end of an enormous shower. The energy of the incident photon has apparently been split by the multiplicative process into an enormous number of very low-energy photons and electrons. This photograph furnishes excellent evidence for the correctness of the shower theory.

definitely identified as having been produced at essentially the same instant. From the early autumn of 1931, when we got our magnet built and began our direct study of the energy of cosmic-ray photographs, we differentiated between two types of showers. Thus, when two or more tracks appear to spring from a common centre we called them "space-associated tracks", but from the first we found photographs on which two or more tracks appeared not to spring from a common centre, and from their identity in sharpness or diffuseness to be clearly produced at practically one and the same instant. Fig. 14 shows one of the earliest of these, taken in the autumn of 1932 and published in March 1933. A difference of a small fraction of a second in time of formation causes a notable difference in the diffuseness of the track. Tracks of the same sharpness we designated as "time-associated tracks". The majority of our showers contained tracks that were associated in both time and space, as is to be expected from the pair theory.

It will be seen from a glance at the photographs of space-associated tracks that in general the particles whose movements they reveal all appear to be moving with very little divergence from a common direction, save as the magnetic field bends them out of this direction. The particles thus reveal clearly in most cases the direction of the incoming cosmic ray, and yield direct ocular evidence for the validity of the law of conservation of momentum in cosmic-ray impacts of this sort. When heavy particles are involved, this evidence is, in some cases at least, wholly wanting (see below, Fig. 18).

Also, the space-associated showers like that shown in

Fig. 7, p. 32, taken at sea level, reveal energies apparently brought to the chamber by a single photon which reaches in some of our photographs the very large energy value of 3 or 4 billion (10^9) e-volts. Also the shower shown in Fig. 15, taken on Pike's Peak (14,000 feet), is estimated from the number and straightness of the individual tracks to have originated from the absorption at or near that point of an incident cosmic ray of the huge energy of 20 billion e-volts. However, the way in which these very many tracks (more than a hundred of them) seem to spring more or less from a common centre is not easy to reconcile with the hypothesis that a single encounter between a photon and a nucleus can give rise to but a single electron pair. It may be necessary to assume that a large number of pairs may *sometimes* be formed at a single encounter, though it will be shown later that if this is true the phenomenon must be a relatively rare one.

By taking a large number of photographs with the Pasadena apparatus on Pike's Peak in the summer of 1935, Anderson and Neddermeyer found the percentage of cosmic-ray photographs showing showers notably greater than at sea level. They measured the increase in the frequency of showers containing 2, 3, 4, 5 and many tracks, and in each case found it much greater than the increase in electroscope readings. This checked entirely the results found by Rossi, Montgomery,* Young,† Woodward,‡ and others who had studied by the so-called "counter

* C. G. and D. D. Montgomery, *Phys. Rev.* **47**, 429 (1935).

† Young, *Phys. Rev.* **49**, 638 (1936).

‡ Woodward, *Phys. Rev.* **49**, 711 (1936). See also Braddock and Gilbert, *Proc. Roy. Soc.* **156**, 570 (1936).

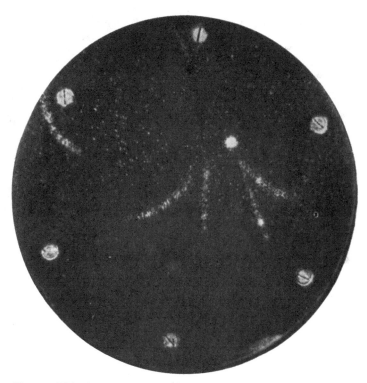

Fig. 14. This photograph shows five, possibly six, time-associated but not space-associated tracks. They were all formed perhaps a second or so before the expansion, so that the ions had had time to diffuse. This kind of a track lends itself well to counting the number of ions formed per cm of length of track.

Fig. 15. An enormous shower taken on Pike's Peak. It is produced by a cosmic ray which had an energy of some twenty billion e-volts, as roughly estimated from the number and curvature of the tracks.

technique" the prevalence of showers at sea level and at higher altitudes. On the other hand, all observers who have compared the prevalence of showers at sea level and at greater depths beneath the top of the atmosphere, Rossi, Auger, Clay, Pickering, etc., either by sinking instruments in the sea or by making measurements in tunnels or beneath reservoirs of water, have in such experiments found the percentage of showers remaining essentially constant. Clearly, then, the component of the cosmic radiation that is most effective in producing showers is more abundant on mountain tops than at sea level, but not particularly more abundant at sea level than at lower depths.

It is also true, as shown by Bowen, Millikan and Neher, that the *average penetrating power* of the rays existing at considerable elevations above sea level (10,000 to 30,000 feet) is very much less than that existing at sea level or below. Also in 1933, by taking up in bombing planes at March Field, to altitudes of 20,000 feet, electroscopes shielded by large thicknesses (up to 16 cm) of lead, iron and aluminum, Millikan and Neher found the *relative absorption* coefficients in these three substances approaching closely, with increasing thickness, a so-called mass-absorption law (absorption proportional to density) which is characteristic of particles having a *range* and entirely different from the absorption law found for cosmic rays in the upper part of the atmosphere. In other words, some change in the character of the radiation both as regards (1) average penetrating power or absorption law, and (2) as regards the facility with which it produces showers, begins to be quite noticeable as one approaches from above the portion of the atmosphere near sea level, and this changed character persists at lower depths.

4. *Distribution of measured energies, and sign of charge among the very high-energy single cosmic-ray tracks.*

As already indicated, about 85 % of all the successful cosmic-ray exposures taken at Pasadena by the random method reveal isolated tracks coming in almost exclusively from above and at angles of less than 60° from the vertical. Some indication of the distribution of energies among these rays can be found from the early Millikan and Cameron experiments on the rate of decrease in ionization in electroscopes sunk to various depths in snow-fed lakes. These showed that about 70 % of the sea-level ionization had disappeared at a depth of 30 feet of water, i.e. at the equivalent of 1 atmosphere below sea level. The energy dissipated in the "ion track" alone in getting vertically through the atmosphere is of the order of 2 billion e-volts,* and evidence will be given in the next lecture that no actual incoming rays of energy less than about 5 billion e-volts can throw any ionizing influences at all down to sea level. From these two facts it may be inferred that about 70 % of all the rays found at sea level have energies under 5 billion e-volts.

In the same way, since Millikan and Cameron, and also Regener, found 97 % of the sea-level ionization gone at a depth of 300 feet, or 10 atmospheres, we may conclude

* This is obtained simply by multiplying the observed number of ions formed per cm of path (taken as 65) at atmospheric pressure by 32 e-volts, the energy required on the average to free an ion in air, by the number of such centimetres in an atmosphere.

that not more than 2 % of the sea-level cosmic rays can have energies over 50 billion e-volts.

Since Millikan and Cameron had proved, as early as 1925, that the absorbing power of water and air are essentially the same per gram of matter traversed, the foregoing considerations explain altogether satisfactorily why 70 % of the tracks found at sea level come in at zenith angles under 60°, since after the zenith angle of 60° is passed the incoming rays must traverse more than 2 atmospheres of air to exert any influence as far down as sea level.

The most direct evidence, however, for the distribution of energies of sea-level cosmic-ray tracks is found in the cloud-chamber measurements. According to the statistical cloud-track studies of Anderson and Neddermeyer, when they discarded all tracks under three hundred million e-volts as too likely to be produced by showers and extra-nuclear encounters, they found that 75 % of their tracks had energies under 4 billion e-volts; and up to energies of 6 billion volts, which is as far as they are able to differentiate unambiguously between particles of different sign, they found *that positives and negatives are present in equal proportions*, though the positives seemed to be just a trifle in excess. These results, reported in full at the International Conference on Physics at London in September 1934, are generally in accord with the findings of Kunze,* Blackett and Brode,† and LePrince-Ringuet and Jean Crussard.‡ The last of these observers, however, have gone

---

* Kunze, *Zeit. f. Phys.* **80**, 559 (1933).
† Blackett and Brode, *Proc. Roy. Soc.* **154**, 573 (1936).
‡ LePrince-Ringuet et Jean Crussard, *C.R.* **204**, 112 (1937).

further than their predecessors and have done the most extensive and most quantitatively dependable work thus far available on these very high-energy particles, for they have had at their disposal the big Paris magnet which gives them a uniform field of 13,000 gausses over a chamber 50 cm across. With this they estimate that they are able to measure single-track energies up to 20 billion e-volts. Of some 400 measured tracks of energies above a billion e-volts they find about 62 % between 1 and 4 billion and 90 % between 1 and 10 billion, thus showing, as is to be expected from the under-water work already cited, how rapidly the frequency of occurrence of higher energies decreases with increasing energy.

But *most important of all is the definite check of the earlier but less certain finding of Anderson, Kunze and Blackett that positives and negatives appear in very nearly equal numbers in these high-energy tracks.* For since we know, from the so-called east-west effect in the equatorial belt (to be discussed in the next lecture), that high-energy positives come into our atmosphere in marked excess over high-energy negatives (at least 25 % and possibly very much more), the failure of this excess to appear at sea level was at first very puzzling. It can now be explained by the assumption that the positives that come in from outside do not get down to sea level at all, but that *these high-energy and highly penetrating positives and negatives that we observe in our cloud chambers are secondaries produced in our atmosphere* in about equal numbers. Indeed there seems to be no escape from this conclusion (see Lecture III).

The slight excess of positives over negatives may perhaps be explained by another significant and altogether new discovery, if further tests confirm its validity, of the aforementioned French observers, namely, *that high-energy negatives are more easily absorbed in lead than are high-energy positives*. If this is attributed, as seems quite natural, to the attraction which the very strong positive charge of a nucleus, like that of lead, exerts upon the passing negative (the positive would be pushed away instead of drawn in), then the effect of this force would of course be quite small with weakly charged nuclei such as those existing in the atoms of the atmosphere. It might then be understandable why a negative in going from its place of origin in the atmosphere to the cloud chamber would be only slightly more absorbed than a positive, while in going from the place of origin through a lead block 14 cm thick, as it had to do in LePrince-Ringuet's and Jean Crussard's experiments, the difference between the absorption of the positives and the negatives would be very marked, thus causing the observed large excess of positives over negatives after such traversals, even though when the medium traversed had been air only a very slight excess had been observed.

5. *The penetrating power of cosmic-ray tracks at sea level.*

We have already put forward some evidence that the majority of the cloud-chamber tracks found at sea level correspond to highly penetrating particles, since they reach down at least with their ionizing effects so far below sea level, but the direct evidence of cloud-chamber work is even better. It will be remembered that the Oppenheimer-Bethe-Heitler theory does not permit electrons, even of 10 billion e-volts of energy, to pass through more than 2 metres of water or 4 cm of lead, the reason being that the energy of the impulse photon produced by the "bremsung" of the electron on entering a sheet of matter is proportional to its incident energy, or that the *coefficient* of absorption is a constant independent of the energy of the incoming electron. Similarly, with the quickly following transformation of the energy of this photon into an electron pair, etc., etc.

In fact, however, when Anderson and Neddermeyer in 1933 made a statistical study at sea level of the penetration of the isolated cosmic-ray tracks through sheets of carbon and lead they obtained the results shown in the following table:*

TABLE 1. Anderson and Neddermeyer's statistical proof of the great penetrating power of isolated sea-level particles

Material and thickness of plate	Number of electron traversals	Number of secondary negatrons	Number of secondary positrons
1·5 cm carbon	810	38	0
1·0 cm lead	397	15	2 pos-neg pairs
1·1 cm lead	176	7	0
0·2 cm lead	267	3	1 pos-neg pair
0·025 cm lead	789	8	0

The upper row shows that they observed 810 traversals of a carbon plate 1·5 cm thick—this is equivalent to $1\frac{1}{2}$ atmospheres—without the formation of a single electron pair so far as the photographs revealed. The second and third lines show similarly that these track-forming particles pass through large thicknesses even of lead without any

* Anderson and Neddermeyer, *International Conference on Physics*, London, 1934.

apparent pair formation. Similarly, LePrince-Ringuet's and Jean Crussard's observations show that more than half of the rays found at sea level of energy from 1 billion to 10 billion e-volts seem to be able to pass through 14 cm of lead. Similarly, too, the counter experiments of Rossi and others show conclusively a penetrating power of the foregoing sort for many of the rays observable at the earth's surface.

These facts are so clear and convincing that there appears to be no possibility of escaping the conclusions (1) that particles of very high energy and of very high penetrating power exist in the cosmic rays *in the neighbourhood of sea level and below*; (2) that these are approximately half positively and half negatively charged; and (3) that their tracks are indistinguishable from electron tracks.

There seemed to be no conclusion possible, therefore, save that the Bethe-Heitler law, so well verified by Lauritsen and his pupils, in the range of energies up to say 15 million e-volts, loses its validity in the very high-energy range. I had myself drawn this conclusion, as had also Bethe and Heitler themselves.

6. *X-particles or mesotrons.*

But within the past three years Anderson and Neddermeyer, by making a statistical study of the loss in energy of *shower* particles, instead of isolated tracks, in going through a plate of platinum or lead, have shown that at least up to about 300 million volts this particle loss in a cm of lead is proportional to the energy of the incident particle, and Blackett and Wilson have just confirmed this result in this range. This is precisely as is required by the Bethe-

Heitler theory. Further, Bowen, Millikan and Neher have quite recently shown, as will be detailed in the next lecture, that a law essentially like the Bethe-Heitler law holds for the absorption in air of electrons having an energy as high as 10 billion e-volts, so that *the assumption of the breakdown of the Bethe-Heitler law at such energies*, at least in the atmosphere, is no longer a possibility. To account, then, for the high penetration which Anderson and Neddermeyer find in their statistical studies of the isolated tracks not apparently associated with showers at all, these authors as early as 1934 (see Table 1 above) felt obliged to assume that the observed highly penetrating tracks are not the tracks of electrons, as they seem to be, but of particles of enough mass so that they do not make radiative collisions, i.e. do not lose the normal amount of energy through radiative collisions. *If these particles are protons, then since they are of both signs the negative proton has been discovered.* This would be an exceedingly important discovery if correct. But there are the best of reasons for thinking that they are *new particles* just like positive and negative electrons, save that their mass is intermediate between that of electrons and that of protons. Hence, for the time being at least, they have been called X-particles.* Anderson has recently,

* The history of this discovery is as follows: In 1934 it was pointed out by Anderson and Neddermeyer (*Report of the International Conference on Physics*, vol. 1, Section on Nature of Cosmic-Ray Particles and footnote, p. 182 (1934)) that there are serious difficulties in identifying the penetrating cosmic-ray particles at sea level with either electrons or protons. Further evidence of a new kind indicating the existence of particles of a new type was reported by them in the *Phys. Rev.* **50**, 270 (1936), captions under Figs. 12 and 13. In a colloquium on 12 November 1936 new evidence showing a difference in penetrability of single particles and shower particles was presented, in which the

however, suggested the very appropriate name mesotron (intermediate particle), and in these lectures this designation will henceforth be used.

The direct evidence, as published in 1936 and 1937 (see footnote), for the existence of these new particles was found in the fact that the loss in energy of electron tracks that are a part of showers in going through a plate of platinum is very high, and such electron tracks are found directly from the photographs to fritter away their energy

conclusion was reached that these data could be understood only in terms of a new type of particle. A brief report of this colloquium was published in *Science*, 20 November 1936, p. 9 of the supplement. Further publication was withheld until a long series of careful measurements had been completed on fifty-five particles which showed that shower particles of a given curvature have an entirely different penetrability from non-shower particles of the same curvature. Further, practically all of these particles had much too small ionizing power to be protons. These results were reported in the *Phys. Rev.* **51**, 884 (1937). At practically the same time as this last publication, Street and Stevenson, Abstract No. 40, *Phys. Rev.* **51**, 1005 (1937), presented data showing the existence of particles less ionizing than protons and more penetrating than electrons could be on the assumption *that the Bethe-Heitler law holds for high enough energies.* They thus confirmed the conclusion reached by Anderson and Neddermeyer in 1934 based on measurements reported to the London conference at that time. That the Bethe-Heitler law does not break down at very high energies, as it had been assumed to do by Millikan, Bethe-Heitler, Oppenheimer and many others, was first shown in the San Antonio-Madras experiments published in *Nature*, **140**, 23 (1937) and the *Phys. Rev.* **52**, 80 (1937). Yukawa (see collected papers from Faculty of Science, Osaka University, vol. II, p. 52 (1935)) for purely theoretical reasons substituted in the equation $m_u = h/\lambda c$ a value of λ equal to the diameter of the nucleus, namely, $\lambda = 5 \times 10^{-12}$ cm, and thus got m_u coming out 2×10^2 electron masses. He then remarks: "As such a quantum with large mass and positive or negative charge has never been found by the experiment, the above theory seems to be on a wrong line."

in producing showers (see Figs. 11 and 12), as they ought to do according to the Bethe-Heitler theory; whereas when Anderson and Neddermeyer choose *isolated tracks*, even of the same curvature as the shower tracks, to make their measurements upon, they find in general that these isolated tracks do not lose much energy in going through their platinum bar and do not make showers. They conclude, therefore, that these penetrating tracks, which are about equally distributed among positives and negatives, cannot be electron tracks, though the only way they are differentiable from them with certainty is through their greater penetrating power and non-shower-forming properties. Figs. 16 and 17 show some of these penetrating and non-shower-producing tracks which may be compared with the tracks of about the same initial curvature in the same magnetic field shown in Figs. 11 and 12. These latter lose most of their energy and produce showers in going through the same platinum bar in which the tracks shown in Figs. 16 and 17 come through with about the same curvature with which they entered, and do not make showers though they do show scattering. These particles that do not make radiative collisions or "bremsstrahlung" must be heavier than electrons, which *do* transform their energy into "bremsstrahlung" just because they are so light and therefore so easily "bremsed".

In any case *a new particle* in addition to the positron, and not a *negative proton*, as most Europeans have thought these penetrating particles to be, *but something still more fundamental, has just entered the world of physics through the study of cosmic rays*. This particle is probably knocked out of the nuclei of the molecules of the air, directly or

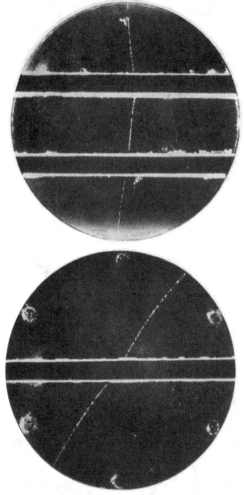

Fig. 16 also shows the tracks of two mesotrons which penetrate in one case two lead bars each a centimetre thick without producing visible showers or changing largely their curvature. Contrast the behaviour of the tracks shown in Figs. 16 and 17 with those in Figs. 11 and 12, which are the tracks of *electrons*.

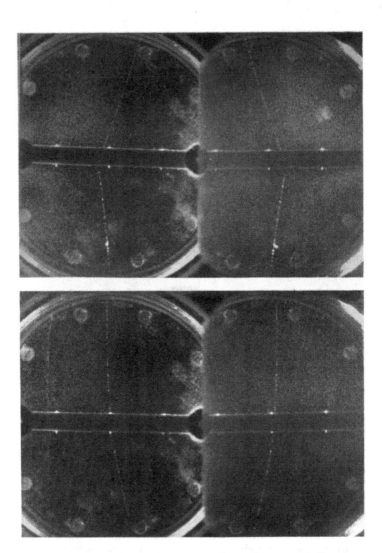

Fig. 17 shows the usual type of isolated cosmic-ray track found at sea level. The upper picture shows the track of a negative particle, the lower that of a positive. Both of these tracks are seen to pass through a platinum bar one centimetre thick without producing any showers or changing much their curvature. They are the tracks of mesotrons.

indirectly, by the incoming photons or electrons. This suggestion requires that there be other types of nuclear absorption possible besides those assumed in the Bethe-Heitler theory, which takes account only of scattered X-radiation, or impulse radiation, produced by incoming electrons and pair formation produced by resulting photons. And indeed one such new form has already been found, as shown in the following experiments, though it does not as yet yield the highly penetrating tracks. It does, however, prove that the process of nuclear absorption contains some other elements than those found in the Bethe-Heitler theory.

7. *The ejection of proton-like particles under cosmic-ray bombardment.*

In one of our earliest photographs taken with the aid of the large magnet we obtained a heavy positively charged track and a light negative track apparently issuing from the same centre, and we assumed and reported in 1931 in a lecture given at the Institut Henri Poincaré in Paris that the absorption of a photon by a nucleus might take place in such a way that a proton-electron pair might be formed. This was before the discovery of the positive electron, but the evidence of this photograph is not changed by that discovery. If this kind of an event is possible at all, it is a rare occurrence relative to the formation of an electron pair. But that particles as heavy as protons can sometimes be ejected from a nucleus by cosmic-ray encounters was brought to light sharply by the cloud-chamber experiments both at sea level (Pasadena) and on Pike's Peak. It is noteworthy that no evidence appears in any

of our photographs that protons, α-rays, or other nuclei
come in from outside and get down to Pasadena or to the
top of Pike's Peak. Protons, however, might not be distinguishable from electrons so long as their energies are
high, but when they are slowed down enough their tracks
should become very easily distinguishable. Our group
has taken 10,000 photographs on Pike's Peak and 100,000
at Pasadena without finding more than a very few tracks
which are clearly those of stopping protons.

But both the Pasadena photographs and those taken on
Pike's Peak do show a certain number of the tracks of
heavy particles which ionize very much as do protons.
These are practically always, however, short tracks which
are clearly due to the explosion of a nucleus through the
re-direction of the energy that has come into it from the
cosmic ray (see Fig. 18). When a nucleus is hit by a photon
the electron tracks always move on in the direction of
the incoming cosmic ray. But *when these heavy proton-like
particles are ejected, their direction of ejection is found in general
to bear no relation to the direction of the incident ray.* Anderson
and Neddermeyer got about 100 such cases out of their
10,000 Pike's Peak photographs. The character of the
tracks shows quite clearly that the incident energy has
merely disintegrated a nucleus, and the heavy constituents
of that nucleus have been thrown out in the disintegration
process.

None of these observed proton energies, save possibly
one somewhat uncertain case, has risen above 100 million
e-volts, so that the foregoing highly penetrating tracks, those
of mesotrons, are not explained by this observed nuclear
disintegration process. However, since proton and electron

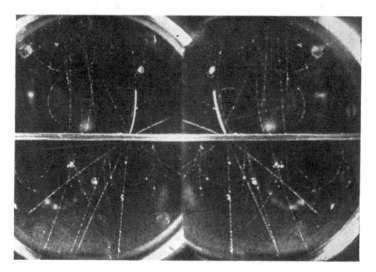

Fig. 18. This photograph, taken on Pike's Peak, shows six tracks all emanating from a common centre in the lead bar. This was clearly the point of collision of about a billion-volt cosmic ray with an atomic nucleus, the disintegration of which apparently resulted in the ejection in all directions of five mesotrons (possibly electrons) and a heavily ionizing proton-like particle moving upward. Such nuclear disintegrations under the influence of cosmic rays are apparently rare events.

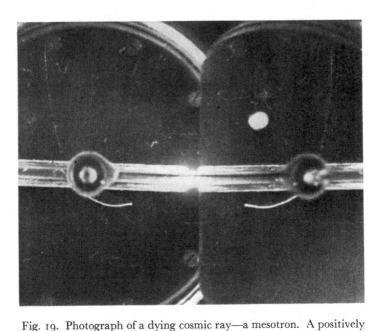

Fig. 19. Photograph of a dying cosmic ray—a mesotron. A positively charged particle of about 220 electron-masses and 10 mev energy passes through the glass walls and copper cylinder of a tube-counter and emerges with an energy of about 0 21 mev. The magnetic field is 7900 gausses. The residual range of the particle after it emerges from the counter is 2·9 cm in the chamber (equivalent to a range of 1·5 cm in standard air). It comes to rest in the gas and may disintegrate by the emission of a positive electron not clearly shown in the photograph. It is clear from the following considerations that the track cannot possibly be due to a particle of either electronic or protonic mass. Above the counter the specific ionization of the particle is too great to permit ascribing it to an electron of the curvature shown. The curvature of the particle above the counter would correspond to that of a proton of 1·4 mev and specific ionization about 7000 ion-pairs/cm, which is at least 30 times greater than the specific ionization exhibited in the photograph. The curvature ($\rho \simeq 3$ cm) of the portion of the track below the counter would correspond to an energy of 7 mev if the track were due to an electron. An electron of this energy would have a specific ionization imperceptibly different from that of a usual high energy particle which produces a thin track, and in addition it would have a range of at least 3000 cm in standard air instead of the 1·5 cm actually observed. Moreover, if the particle had electronic mass and emerged from the counter with a velocity such that its specific ionization were great enough to correspond to that exhibited on the photograph, its residual range (in standard air) would be less than 0·05 cm instead of the 1·5 cm observed. A proton of the curvature of the track below the counter would have an energy of only 25,000 ev and a range in standard air of less than 0·02 cm.

tracks above 600 million e-volts are not distinguishable, cloud-chamber experiments on particles of energies of say a billion volts or more do not yield definite and conclusive evidence that such highly penetrating particles may not be protons; and that they *are* such has been the view, as above indicated, of most European students of cosmic-ray phenomena. The two points that have been urged thus far against this conclusion are (1) that with all the cloud-chamber photographs taken we have failed to bring to light the ends of proton paths, that is proton tracks of energies between 100 and 600 million e-volts, where the ionization along the track should be so much greater than that in an electron track, that the difference could not be mistaken; and (2) that the existence of nearly as many negative as positive highly penetrating tracks means definitely on this hypothesis the existence of negative as well as positive protons, and these have never appeared before in any sort of phenomena. The last difficulty is not serious; indeed, it would be welcomed by the theorist as an altogether acceptable discovery should it prove to be one. The first difficulty, too, can be circumvented by making suitable assumptions as to the mechanism by which, when its energy falls below a certain value, the proton becomes unobservable by transformation into a neutron, for example, or otherwise disappears.

Perfectly direct and unambiguous photographic evidence, however, for the existence of a particle of mass intermediate between that of the electron and the proton has just been obtained from a photograph taken in June 1938 by Nedder-meyer and Anderson. So as to catch particles at the end of their range where the ionization is so unambiguous an index of the mass of singly charged particles, the foregoing experimenters placed one of the two counter releases above the chamber and instead of placing the other counter below the chamber, as is usual, they placed it in the middle of the chamber itself, for clearly the only hope of catching the track of a mesotron at the very end of its range is to have the lower counter within the cloud chamber, so that the particle that is to actuate it after actuating the upper one need not, while within the chamber, possess the same high energy required to go through the lower wall of the chamber in order to actuate the lower counter.

Fig. 19 shows the glass-enclosed counter with a thin copper cylinder within it in the middle of the cloud chamber. It also shows above the counter the track of a particle which, if it is a proton, has a curvature corresponding to an energy of 1·4 million e-volts. But the ionization produced by a proton of this energy is about 7000 ion-pair per cm which is at least thirty times larger than is here found. Again, a proton of the curvature of the track shown below the counter could have an energy of but 25,000 e-volts, and such a proton would have a range in standard air of but 0·02 cm, while the actual range is 1·5 cm. It cannot therefore be a proton but must, in order to have this range of 1·5 cm below the counter, be much lighter than a proton. Again, it cannot be an electron, for an electron ionizing as strongly as this does below the counter would have a range of only 0·05 cm instead of 1·5 cm. To have the observed range combined with the observed curvature the particle must be considerably more massive than is an electron. The photograph provides several methods of computing that mass.

It comes out about 220 electron masses. How many different kinds of mass these mesotrons can have future research alone can determine.*

The discovery of masses intermediate between that of the electron and the proton seems at first very upsetting to our heretofore established knowledge as to the constitution of the atom. Why have we not run across these mesotrons before? The answer probably is that our established knowledge is all right for the range of energies heretofore at our disposal and that under the influence of the stupendous energies involved in cosmic rays matter itself takes on new properties. Other evidence bearing upon this important question will be presented in the next lecture.

It may be profitable to summarize as follows a few of the unquestionable facts brought to light by cloud-chamber photographs of cosmic rays:

1. The earth is being bombarded continuously by superpower particles, photons, electrons, or both, the energy of which rises to directly measured values of at least 20 billion (10^9) e-volts, at least a thousand times the values of any particle energies existing on earth. A hundred or more such particles shoot each minute through the head of every person living on the earth.

2. The great majority of these shots that pass through a cloud chamber at sea level consist of isolated, singly

* Corsan and Brode, *Phys. Rev.* **53**, 773 (1938), and very recently Nishina Takeuchi and Ichímíya, *Phys. Rev.* **55**, 585 (1939), also have taken some excellent cloud-chamber photographs, from which they estimate the mass of the penetrating particles to be about 200 electron-masses.

charged particles endowed with an average energy of a billion e-volts or more, and capable of passing in a nearly straight line through as much as 10 cm of lead.

3. These highly penetrating particles all carry the unit electrical charge and are nearly equally divided in sign of charge between positives and negatives, the positives being apparently a little in excess.

4. Photons carrying energies as high as several billion e-volts are found in cloud chambers at sea level, where they make their presence manifest by the creation within lead strips placed inside the chambers of showers consisting of both positive and negative electrons.

5. These shower particles are very much more absorbable than the particles described in 2 and 3, their loss in energy in going through lead plates being nearly proportional to the incident energy, i.e. they possess an absorption *coefficient* that is independent of energy and otherwise obey the so-called Bethe-Heitler law of electron absorption.

6. Showers increase much more rapidly than electroscope readings in going from sea level to the top of high mountains, but below sea level their abundance is nearly proportional to electroscope readings.

7. The character of shower photographs lends support to the Bethe-Heitler theory of shower formation, namely, that an incoming electron of high energy first transforms that energy largely into an "impulse photon" (bremsstrahlung), that this quickly produces an electron pair, that each electron of this pair repeats the process, etc., until the energy is all degraded into a very large number of lower and lower energy photons and electrons.

8. The foregoing is only one, though quite definitely

the chief, of the possible modes of absorption of the energy of incoming electrons and photons. The incident energy sometimes disintegrates nuclei with the ejection of heavy but not greatly penetrating particles.

9. The incoming energy may throw out of the nucleus very penetrating particles (+ and −) which do not transform their energy into impulse radiation and hence do carry the energy down to considerable depths below sea level. These penetrating rays or mesotrons may in their turn, directly or indirectly, produce *showers* even at sea level and below.

The Earth's Magnetic Field and Cosmic-Ray Energies

It is a matter of more than mere academic or scientific interest that the earth is a great magnet with its magnetic lines of force reaching out into space for a distance of two or three earth's diameters—ten thousand miles and more—as shown in Fig. 20, though the earth's atmosphere extends up for only a few hundred miles at most. For the inhabitant of the Americas, whose dwelling-place was discovered with the aid of the compass, it does not take long to find the answer to the question as to the usefulness of studying the earth's magnetism, for the practical development of the world's commerce was made possible by this apparently useless discovery. But the cosmic rays have found a new use for the earth's magnetic field.

1. *Early experiments on the latitude effect in cosmic rays.*

I have violated the historical sequence of events in explaining how the energies of charged-particle bullets have been measured with the aid of powerful artificial magnetic fields before showing how the earth's magnetic field was first put to this new use. Störmer of Norway in his studies on the aurora borealis had long been engaged in working out with great skill and precision the effects of the earth's magnetic field upon low-energy electrons

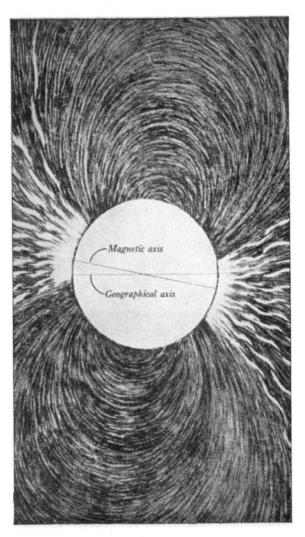

Fig. 20. The earth's magnetic field. The figure shows the shape of the field and the displacement of the magnetic from the geographical poles. The earth's atmosphere extends but a few hundred miles, at most, from the surface, but the magnetic field should be appreciable to a distance of 12,000 miles. At 4000 miles its intensity is one-eighth as much as at the surface. (From an original drawing by W. J. Peters of the staff of terrestrial magnetism, Carnegie Institution of Washington.)

emitted by the sun and absorbed in the outermost regions of our atmosphere.

As soon as Millikan and Cameron in their 1925 experiments on the discharging effects of electroscopes in snow-fed lakes had demonstrated to their own satisfaction the enormous penetrating power and the extra-terrestrial origin of the cosmic rays, they planned a voyage to the west coast of South America primarily to look for the effects of the earth's magnetic field on incoming cosmic-ray electrons. For even if the rays in their origin were of the γ-ray or photon type, as seemed quite probable, yet since photons only make their presence known by the electrons which they detach from the matter through which they pass, it was inevitable, in the case the rays originate in portions of the universe in which matter is present even in small density, that the rays which get out into interstellar space and ultimately enter the earth consist in part, at least, of high-energy electrons.

It is at once obvious that such a high-energy electron approaching the earth at the equator, where its motion is for the most part at right angles to the earth's magnetic lines of force which, though thousands of miles out in space, run at the equator parallel to the earth's axis (see Fig. 20), will tend to be bent around by that field and deflected out into space again like a comet coming into the gravitational field of the sun, while the electrons which approach the earth near the poles, moving mainly along the earth's magnetic lines, will have no difficulty in reaching the surface, since electric currents moving parallel to the lines of a magnetic field are not influenced at all by it. We anticipated, therefore, a diminution in the cosmic-ray intensity as we moved from Los Angeles into the equatorial belt and over the magnetic equator. We discussed this expectation before the American Physical Society meeting in Pasadena in the autumn of 1925, spent the winter building and testing three new electroscopes especially for the voyage to South America, and in the summer of 1926 embarked for Mollendo, Peru, and took continuous readings in the radio room on the top deck night and day on two of our electroscopes (the third proved too responsive to the ship's vibrations to make dependable readings possible) during the sixteen days of the voyage to Mollendo. *Unfortunately, however, we were able to take readings only from the second day out.*

At Balboa we took all three of our electroscopes out into a row-boat in the harbour and took readings there on all of them for twenty-four hours so as to make sure that we were free from any possible radioactivity of the materials around our position on the top deck of the ship.

Our cosmic-ray electroscopes were the relatively crude ones of that day and we estimated that the error in the relative cosmic-ray intensities in the different latitudes traversed might be as much as 6%. So, in our presentation in September 1927 of the results of this voyage in a public address at the meeting of the British Association for the Advancement of Science in Leeds* we wrote as follows: "If then there are any geographical differences in the altitude ionization curve" at sea level or above "they are beyond the limits of our present observational technique", estimated in this and other articles† at 6%.

* *Nature*, **121**, 21 (1928).   † *Phys. Rev.* **31**, 170 (1928).

As a matter of fact there is a decrease of 7 % in sea-level cosmic-ray intensity in going from Los Angeles to Mollendo, as we proved by most careful and repeated experiments in 1932 and 1933, so that we were wrong in not bringing any effect to light in 1926. We should have done so even with our insensitive electroscopes of that day had we not had the misfortune to have started our voyage just at the critical latitude where the reduction in intensity in going south begins to set in strongly, and then had been unable to get our instruments set up and working until the second day out when we were over 500 miles south of that latitude —so far south, indeed, that more than half of the total drop in intensity in going from Los Angeles to the equator had already taken place. The change from that point on was, in fact, so small as to be quite beyond the sensitivity of our instruments. We missed the sea-level effect on this trip, then, just because of the necessary delay at the very start of the voyage in getting the instruments going. On the mountains of Bolivia we did, indeed, get and record, in our published graph, readings 11 % lower* than we had obtained in the northern latitudes with the same instruments, but both on account of our uncertainties as to the shielding of adjacent mountains and possible electroscope leaks we explained these differences away and drew one mean curve through all our northern and southern hemisphere points, treating the divergences as observational uncertainties. Actually if, instead of doing this, we now draw two smooth curves through the recorded points we obtain practically the same percentage differences shown

* See particularly p. 166. The Muir Lake point at 7·85 m. It is 11 % above the graph, just as we now know it should be.

in the two lower curves of Fig. 21. We did not do this because we trusted our much more prolonged and dependable sea-level readings, and for several years thereafter in order to account for the apparent absence of a latitude effect we made erroneous hypotheses as to the exclusively photonic nature of the incoming rays. The actual bringing of the looked for effect to light came about thus:

In an article dated 17 December 1927 Prof. J. Clay made a qualitative report* of a latitude decrease between Amsterdam and Batavia; in another article dated November 1928† he gave the numerical value of the cosmic-ray ionization as 1·49 at Leiden and 0·82 at Singapore, a very large percentage of change, and in a later article dated 30 September 1930‡ he modified these ionizations to 1·24 at Amsterdam and 0·93 at Singapore, still a large difference. Furthermore, a long series of observations taken in 1928 on the *Carnegie* under the direction of Dr Gish in equatorial, temperate, and northern latitudes in the Pacific were reported to me privately by him, since I had helped him in calibrating his instruments, and seemed, with some uncertainty, to indicate a 5 or 6% lower equatorial value than was found in the more northerly latitudes. Stimulated by the diversity in these reports, as well as by the foregoing theoretical considerations which seemed to demand at least some electronic component in the cosmic rays, and having by this time developed pressure electroscopes a dozen times more sensitive than our earlier ones, in the summer of 1930 I made, with a sensitive pressure electroscope, a careful series of readings at Pasa-

* *Proc. Roy. Acad., Amsterdam*, **30**, 1115 (1927).
† *Ibid.* **31**, 1091 (1928).   ‡ *Ibid.* **33**, 711 (1930).

dena (geog lat 34°) of some weeks' duration, then transported the same electroscope to Churchill, Manitoba (geog lat 59°) and, keeping the room in the small wooden house in which I worked at Churchill at practically the same mean temperature as at Pasadena, I found the mean value of a week's reading taken there the same as at Pasadena with an error which I estimated could not in this case be more than 1 %. I had thus gone south at sea level from Pasadena to beyond the equator and north to a point reasonably close to the north magnetic pole without bringing to light any effect of the earth's magnetic field on incoming electrons at all, though the southerly measurements had never been claimed to be better than 6 % in accuracy. The measurements from Pasadena northward, however, had been done with such care and represented means taken over so long periods of time that I was quite confident as to their accuracy. Also, they were not in disagreement with the results obtained in the same summer (1930) by Bothe and Kolhörster,[*] who had started from northern Europe and taken continuous readings as they voyaged northward between geographical latitudes 53° N and 81° N without noticing a sea-level latitude effect.

Further, in 1930 I had asked Prof. Paul Epstein to work out quantitatively the expected effect of the earth's magnetic field on incoming electrons of very high energy, such as a billion e-volts, and he had obtained the result published briefly in December 1930[†] that electrons of that energy would come in only north of a certain critical latitude λ_2, being completely eliminated south of

[*] Bothe and Kolhörster, *Berliner Berichte*, p. 450 (1930).
[†] Epstein, *Proc. Nat. Acad. Sci.* **16**, 658 (1930).

it.* The analysis was carried further in 1933 by LeMaitre and Vallarta, who showed that the intensity of these electrons increases north of λ_2 in a zone extending to a second critical latitude λ_1. Within the polar cap north of λ_1 the intensity is uniform. These two critical latitudes, λ_1 and λ_2, would of course move south if the energy of the incoming electrons were assumed to be greater. Also, if the energies were variable over a considerable range, the uniform sea-level polar cap would have to set in at the value of the incoming electron energy which was just sufficient to enable the effects of the incoming electrons to reach down to sea level through the resistance offered by the atmosphere. For lower energies than this the polar cap would begin farther north, but because of the absorption of the atmosphere it would of course be necessary to go to altitudes above sea level to bring it to light, since the energy required to get through the earth's magnetic field would here be smaller than the energy required to reach down to sea level. These conclusions from Epstein's analysis, familiar to all the workers at the Norman Bridge Laboratory from the autumn of 1930, explained why, even if there were some electron component in the incoming cosmic rays, I had found such surprising and precise constancy in sea-level cosmic-ray intensities everywhere between the latitude of Pasadena (mag lat 41° 30′) and regions close to the north magnetic pole. Guided, then, by these computations of Epstein's, which showed that it ought to be much easier to bring a latitude effect on incoming electrons to light at high altitudes than at

* Implicitly these results were contained in the earlier work of Störmer.

low, in the autumn of 1931 I made application to the Carnegie Corporation of New York for funds to make a new survey, this time at the highest altitudes that could be reached by airplanes, of the effect of the earth's magnetic field on incoming electrons.*

This series of observations was made in the summer of 1932, when Bowen, Millikan and Neher took new vibration-free, self-recording Neher electroscopes up in airplane flights to altitudes of 21,000 feet in the five different latitudes of Cormorant Lake, Manitoba (mag lat 63° N), Spokane, Washington State, U.S.A. (mag lat 54° N), March Field, near Pasadena (mag lat 41° N), Panama (mag lat 20° N), and Peru (mag lat 4° S).

Before these observations had been completed, Prof. A. H. Compton† had begun a re-survey of cosmic-ray intensities conducted on the earth's surface, and had reported that he and his collaborators had checked Prof. Clay in finding a sea-level equatorial dip in crossing the equator, in his case on the west coast of South America and in the Pacific. He also reported a considerably larger equatorial dip in his observations on mountains. Also, in returning from Peru to New York in December 1932, the foregoing new Neher electroscope recorded a 6·8 % equatorial dip between Mollendo and New York. This brought all sea-level observers into agreement in the essential point of finding a small equatorial dip in the cosmic-ray intensities no matter from what point in the

* These and the foregoing reasons for them were discussed quite fully in a lecture given at the Institut Henri Poincaré in Paris on 2 November 1931 (see *Annales de L'Institut Henri Poincaré*, pp. 450, 451 (1933).

† Compton, *Phys. Rev.* **41**, 111 (1932); **43**, 387 (1933).

north temperate zone one starts south across the equator. The differences in the numerical values of this dip, at first surprisingly large, are rapidly becoming narrowed down and will presently be no longer a subject of uncertainty. They are not now of any fundamental interest save as another illustration of how inevitably the scientific method of approach reveals the difference between a right result and a wrong result, or even a correct theory and an incorrect one, and renders further dispute impossible among men who have learned to be guided by knowledge rather than by preconception and prejudice.

2. *Some results of the airplane latitude survey of the latitude effect in cosmic rays.*

The 1932 airplane latitude survey of Bowen, Millikan and Neher* brought to light, however, some new facts, the understanding of which is of fundamental importance for our later attempts at the general interpretation of cosmic-ray phenomena. Thus, although in new observations made in the summer of 1932 Millikan and Neher got no measurable difference at all in sea-level intensities in going by sea from Pasadena to Seattle and Vancouver (see Table 2, in which we have estimated 1 % as the limit of our observational uncertainty†), yet the airplane flights

* First reported briefly in December 1932. See *Science*, **43**, 661–9; also *Phys. Rev.* **44**, 246 (1933), and *International Conference on Physics*, I, Nuclear Physics, p. 210 (London, 1934).

† A new set of careful sea-level observations by Compton and Turner between Vancouver and Sydney, analysed and summarized by Thompson in *Phys. Rev.* **52**, 141, is in substantial agreement with all our sea-level measurements within the limits of our estimated uncertainty, namely, about 1 % between Vancouver and Pasadena, with a very rapid decrease in intensity amounting to more than 3 % in going 500 miles south of mag lat 41°.

taken at March Field (mag lat 41° N, near Pasadena) and Spokane (mag lat 54° N, nearly the same as Seattle) up to 21,000 feet (4·5 metres of water from the top) show

TABLE 2. Five-day study of sea-level variations of cosmic-ray intensities between Victoria, B.C. and Los Angeles.‡

	Electroscope No. 1—Unshielded (ions cc per sec)			Electroscope No. 2—Shielded (11 cm Pb)			
	Lat	4 hour means	12 hour means	Lat	Long	Mag Lat	12 hour means
In Seattle Harbour	47·5	39·58	39·58*	47·5	122 W	53·5 N	24·60
In and near Victoria				48·3	123 W	54·3 N	24·71
	48·4	39·45	39·45	46·5	124 W	52·5 N	24·65
At Sea	47·22	39·53	39·53	44·0	124 W	50·0 N	24·78
,,	44·4	39·67	39·67	41·0	124 W	47·0 N	24·49
,,	41·6	39·69	39·69	37·5	122 W	43·5 N	24·79†
,,	39·0	39·34	39·33	36·0	122 W	42·0 N	24·57
In San Francisco Harbour	—	39·82	39·82	34·0	119 W	40·6 N	24·45†
,, ,,	—	39·61	39·61			Mean	24·63
At Sea	36·85	39·65	39·65				
,,	34·8	39·63	39·65				
	Mean	39·60					

* 8 hour average.   † 9 hour average.

unambiguously (Fig. 21) a difference increasing with increasing altitude, as Epstein's analysis predicted,* and a similar and very much larger difference between Panama and either March Field or Spokane. These differences are

‡ When I first reported these results in the autumn of 1932 I expressed surprise at a 11% difference at an altitude of 21,000 feet between Spokane and March Field. I expected such differences *farther north*, thinking the evidence at that time good that the resistance of the atmosphere to an incoming electron was about a billion e-volts instead of the 5 or 6 billion required by Epstein's analysis plus my failure to find any increase in cosmic-ray sea-level intensity north of mag lat 41° 30′. I discussed this point rather fully in December 1932 (see *Phys. Rev.* **43**, end of p. 668).

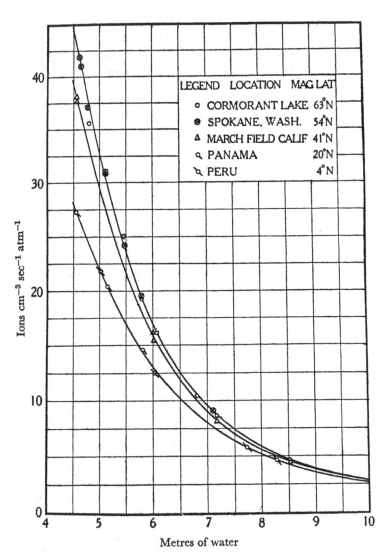

LEGEND   LOCATION   MAG LAT
o CORMORANT LAKE 63°N
⊕ SPOKANE, WASH. 54°N
△ MARCH FIELD CALIF 41°N
⌕ PANAMA   20°N
⌕ PERU   4°N

Ions cm⁻³ sec⁻¹ atm⁻¹

Metres of water

Fig. 21.

also in reasonably good agreement with the mountain-peak observations made by both Compton and ourselves in Peru and the United States, so that there is no question at all about their general correctness.

But by far the most significant results of the airplane survey not obtainable as yet from measurements by other observers are (1) the absence of any difference at all between the depth-ionization curves up to altitudes of 21,000 feet taken at Spokane and at Cormorant Lake, and (2) the similar absence of any difference between the curves taken at Panama and in Peru. The last of these results is of the lesser importance. It merely checks nicely, by the more sensitive high-altitude measurements, the result appearing in all our sea-level readings from 1926 on that there is no appreciable latitude effect between Panama and Peru, a fact consistent with the anticipated very slow rate of change of the earth's magnetic field in the equatorial belt. The first of the foregoing results, how-ever, will repay more careful consideration.

It clearly means that Epstein's polar cap of uniform cosmic-ray intensities exists at an altitude of 21,000 feet down as far as to the latitude of Spokane in precisely the same way as my Pasadena-Churchill observations showed that the Epstein sea-level polar cap of uniform cosmic-ray intensities extends down to the latitude of Pasadena. So far as sea-level effects are concerned, immediately south of Pasadena, as shown by our very careful and prolonged sea-level measurements, as well as by measurements published by Compton and others, a critical latitude is reached at about 41° N at which in going south a reduction in the intensity of the sea-level rays begins to set in, which

reduction reaches a value of some 7 % by the time Panama
is reached, after which the intensity is substantially constant down to the magnetic equator. The beginning of a
reduction at about magnetic latitude 41° means that the
electrons which at this latitude are just able to get through
the blocking effect of the earth's magnetic field have also
just enough energy in some way, direct or indirect, to
throw their influences down to sea level and produce
ionization there. No electrons of lower energy than these
can throw their influences down as far as to sea level, so
that the effects of all the lower energy electrons that can
get through the blocking effect of the earth's magnetic
field north of about latitude 41° are absorbed somewhere
in the portion of the atmosphere above sea level. The
electrons that have the energy both to get through the
blocking effect of the earth's magnetic field and to throw
their influences down to sea level of course get through
both resistances north of 41° just as they do at 41°, and
thus produce the polar cap of uniform intensity predicted
by Epstein's analysis of 1930 and worked out with much
greater elaboration by LeMaitre and Vallarta in 1933*
and later. According to most of the experimental work
that has been done in this region, this *cosmic-ray shelf*, or
discontinuity, that sets in at about 41°, near Los Angeles,
is reasonably sharply marked.† According to Lemaitre

* LeMaitre and Vallarta, *Phys. Rev.* **43**, 87 (1933).

† Dr Neher and I have just discovered by repeating six times in the
years 1938 and 1939 the voyage between Los Angeles and Vancouver
that it is only in summer that there is no difference in sea-level intensity between these points. In winter the sea-level intensity at
Vancouver seems to be some two or three per cent higher than at
Los Angeles, while at the latter point the winter and summer inten-

and Vallarta the mean energy of the electrons that begin to be cut out in considerable numbers by the blocking effect of the field as one moves through latitude 41° toward the magnetic equator is about 5 or 6 billion e-volts, so that the *range* of the influence of say 5·5 billion e-volts may be taken as 1 atmosphere.*

By the same method, the mean energy of the electrons that begin to be cut out in considerable numbers by the blocking effect of the field at the latitude of Spokane (54° N) is 2·3 billion e-volts. The significance of the fact that the depth-ionization curve at Spokane is the same as at Cormorant Lake is found, then, in the proof that it furnishes that just as incoming electrons of an energy of 5·6 billion volts or more create a polar cap of constant intensity at sea level which extends down to the neighbourhood of latitude 41°, so incoming electrons of energy of 2·3 billion e-volts create at 21,000 feet a polar cap of constant intensity which extends down to about the latitude of Spokane.† Another way to state this is that

sities are the same. This kind of a seasonal effect, or "atmospheric-temperature" effect, has been reported earlier by Corlin, Hess, Compton and others. We have tried without success to find a relation between this seasonal effect and the influence on cosmic-ray intensities of magnetic storms as already reported by Forbush (*Terr. Magn.* **42**, 1 (1937)), by Hess and Demmelmair (*Sitz. d. Akad. d. Wiss. in Wien,* **147**, 89 (1938)), and by Clay and Bruins (*Physica,* **5**, 111 (1938)).

* From a study of Lemaitre and Vallarta's curves I have usually taken this critical energy as 6 billion e-volts. Dr Neher, after very much more careful study, using the very elaborate results of Lemaitre and Vallarta, places the most probable energy at 5·6 e-volts.

† So far as these experiments are concerned, the polar cap of constant intensity at 21,000 feet might extend a few degrees south of Spokane, though not many, and in fact experiments to be later discussed fix it close to that latitude.

the range of the influence of 2·3 billion volt incoming electrons is about 4·5 metres of water, or 0·45 of an atmosphere. If the range is proportional to energy, then the range for 1 atmosphere would come out $\frac{2·3}{0·45} = 5·1$ billion e-volts, quite close to where we found it from the data at latitude 41°.

The foregoing considerations appear quite definitely to justify the conclusion that at least in the range of energies involved the distance through the atmosphere to which the influences arising from incoming charged particles can penetrate are proportional to the incident energies, or at least vary *not less rapidly* than the first power of those energies. This is a property characteristic of particles possessing "*a range*". α-particles or β-particles passing through gases lose energy in this way, as do any particles which fritter away their energy uniformly along their paths, for example, by making a more or less constant number of ions per centimetre of ion track, such as one sees in a Wilson cloud chamber. In general, particles that lose energy in that way follow approximately a so-called "mass-absorption law", which means that in going through media of different densities the loss of energy per centimetre of path traversed by the moving particle is proportional to the density of the medium traversed. The reason that low-energy charged-particle rays in general follow more or less perfectly this law is that the number of electrons they meet in going a centimetre in different media is nearly proportional to the density of the medium. But the law of absorption concerning which we are talking is just the law which we saw in the last lecture seems to be followed by the cosmic rays when we experiment upon them in the neighbourhood of sea level. Thus Millikan and Neher took up in army bombers at March Field electroscopes surrounded by varying thicknesses up to 16 cm of aluminum, iron and lead, and found the relative absorptions in these substances approaching with increasing thickness the mass absorption law, as Hoffman and Steinke and their pupils* had found earlier. This is the law, too, that the great majority of the isolated cosmic-ray ion tracks which we find in Wilson cloud chambers at sea level (see Table 1 on p. 44 of the preceding lecture) seem to be following.

In a word, then, the two polar caps of uniform cosmic-ray intensities, the one brought to light by a sea-level latitude survey and terminating just south of Pasadena, the other brought to light by the high-altitude airplane survey at 21,000 feet and terminating in the general latitude of Spokane, are both entirely in accord with a considerable group of other properties possessed by the penetrating component of the cosmic rays. To reconcile these properties with still other properties has not been easy, but the effort has led to new and important knowledge.

3. *The longitude effect in cosmic rays.*

Although many people from 1925 on were expecting and looking for a latitude effect in cosmic rays, nobody apparently ever thought of looking for a longitude effect before it was discovered in January 1934, and it is a commentary on the accuracy of sea-level cosmic-ray measure-

* See, for example, Tielsch, *Zeit. f. Phys.* **92**, 589 (1934).

ments up to that time that it was missed so long, for it is not at all a minute effect. It was discovered quite independently by Clay and by Millikan and Neher, but to Clay goes the priority of publication. This was made in a brief paragraph in the March number of the Dutch journal, *Physica*, 1934, as a result of differences which he observed in the equatorial dip in two voyages made in 1934 between Amsterdam and Batavia, one through the Suez Canal, the other around the Cape of Good Hope. Without any knowledge of Clay's work, we made our first public announcement at the meeting both of the National Academy of Sciences and the Physical Society in April 1934.

From our point of view and from the point of view emphasized in the first of these lectures, this discovery was important because it showed that another element in the interpretation which we had thus far placed on cosmic-ray phenomena was wrong. The simplest way to account for the non-field-sensitive part of the cosmic-ray effects was to assume that it was practically all due to incoming photons, and this was the view that I had consistently advocated up to this time. The discovery of the longitude effect in the equatorial belt itself showed two things: *first*, that the earth's magnetic field, even at great distances above the surface, is not symmetrical with respect to an axis passing through the centre of the earth but has a decided lop-sidedness; and, *secondly*, that at least enough electrons to account for the longitude effect must enter the earth with so great an energy as to be able to break through the blocking effect of the earth's field even in the equatorial belt; in other words, a fraction at least of the non-field-sensitive portion of the cosmic rays, as determined by the latitude effects, must be electronic in nature. This was a fundamental point.

In our case it was not a discovery made all at once. We had been collecting the data for it since the autumn of 1932. The sequence of events leading up to it had to do with the development of accuracy in our sea-level measurements. By two trips on different ships in December 1932 and January 1933 we had fixed the equatorial dip between Los Angeles and Peru at $7 \pm 1 \%$ as measured by a sensitive "Neher electroscope" surrounded by a 10 cm lead shield. Other observers were at this time reporting it as 16% between these same two localities. In August 1933 we placed this same self-recording electroscope within the same shield in the room of Captain Cullen on the Dollar line steamer *President Garfield* and sent it on a three-months' voyage around the world, touching at Honolulu, Kobe, Shanghai, Singapore, the Suez Canal, Genoa and New York. Captain Cullen had only to have the clock wound every day or two. The film, sufficiently long to last through a three-months' voyage, was developed after the return of the ship to Los Angeles and the film to the laboratory. Figs 22–26 show the Neher vibration-free, self-recording electroscope as sent upon this and ten other voyages, and Fig. 27 a section of the developed film. The accuracy with which the rate of discharge during a given hour can be determined is limited only by the accuracy of measurement of slope of each of these discharge lines. The mean of these slopes for a 24 hour period gives the mean cosmic-ray intensity during that day at the place at which the electroscope is found.

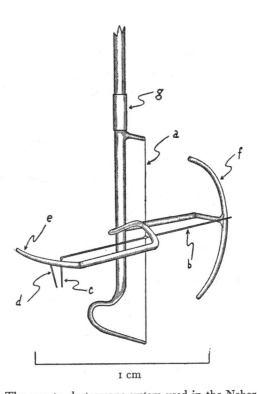

1 cm

Fig. 22. The quartz electroscope system used in the Neher tilt-and-vibration-free cosmic-ray instrument. The system is made entirely of fused quartz. Everything below the platinum sleeve g, on which the charging arm makes contact, is coated with gold by evaporation. The 5μ fibre a is stretched until its length is increased about one percent. The 30μ movable cross-arm b is bent at right angles at one end, where it is drawn down to a thickness of about 10μ. The image of c cast on the recording film by a lens giving a magnification of 10 has then a convenient width on the film. The short piece of fibre d serves as a fiducial mark and it, together with the part e (which is bent into the arc of a circle with the torsion fibre as a centre) and the stop f, combines to give a linear scale over practically the whole range of discharge. A twist of about 30° is placed permanently in the torsion fibre so that no motion of the movable arm takes place until about 250 volts are reached and then the full deflection of 2 mm results for the next 75 volts. The movable part b is balanced by cutting off one end until a tilt of 90° causes less than 0·005 mm actual motion of c. Besides being free from tilt, the system, because of the large ratio of strength to weight, is quite insensitive to vibration.

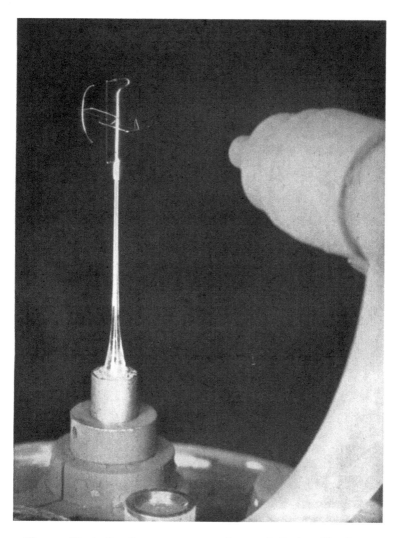

Fig. 23. Illustrating the quartz system when ready for installing into the ionization chamber. The lens mounting is supported by a rigid arm which is securely fastened to the same piece that holds the electroscope system.

Fig. 24. The instrument used in this survey is usually shielded with lead and is placed in the box when used in most airplane flights.

Fig. 25. The camera will take a one-hundred foot reel of 35 mm motion picture film which is driven at a constant rate past the slit by a power clock. Changeable gears allow various rates of film speeds to be used, depending on the expected ionization.

Fig. 26. An 10 cm lead shield protected by a one-half inch shell of cast iron has been used for most of the sea level survey to insure freedom from local variations in radiation.

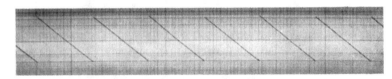

Fig. 27. Showing the type of record obtained at sea level in this world survey. Two of the horizontal lines give barometric and temperature records.

This particular film came back early in December 1933, and by early January 1934 it had been measured and the result found that while the equatorial dip in going from Los Angeles to Mollendo was but $7 \pm 1\%$, in going to Singapore via Kobe and Shanghai it was $12 \pm 1\%$, or nearly 1·8 times as large.

To see whether we would obtain an intermediate value in crossing the equator at a point intermediate between Mollendo and Singapore, in the summer of 1934 we installed two electroscopes, one unshielded and one shielded with 10 cm of lead, in the room of First Officer Graham of the Matson Line steamer. *Monterey*, plying between Los Angeles and Melbourne via Honolulu, and obtained a $10 \cdot 1 \pm 1\%$ equatorial dip on both electroscopes.* This result may be compared with the dip of $10 \cdot 2\%$ just published as the mean dip found in traversing this same track across the equatorial belt in eleven voyages between Vancouver and Sydney made by instruments provided by Drs Compton and Turner.† Also, our value of the equatorial dip found on the west coast of South America may be compared with Clay's recent value‡ in the same area of 6% with lead shields and 8% without shields. Fig. 28 gives in one graph the results of sending in this way Neher instruments on a number of world-encircling voyages, and Fig. 29 gives the combined results of our

* Theoretically there should be some difference between the percentage of equatorial dip obtained with a shielded and unshielded electroscope, but in the three voyages in which we have used unshielded instruments the difference, if it exists, has been inside the limits of our observational uncertainties.

† *Phys. Rev.* **52**, 141 (1937). See also footnote on p. 64 of this book.

‡ Clay, Bruins and Wiersma, *Physica*, **3**, 746 (1936).

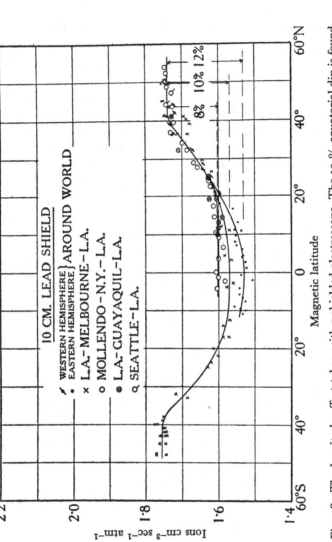

Fig. 28. The longitude effect taken with a shielded electroscope. The 12 % equatorial dip is found in the neighbourhood of Singapore (eastern hemisphere). The 10 % equatorial dip is found in going from Los Angeles to Melbourne. The 8 % equatorial dip is found in going in summer from Seattle or New York to Guayaquil or Mollendo.

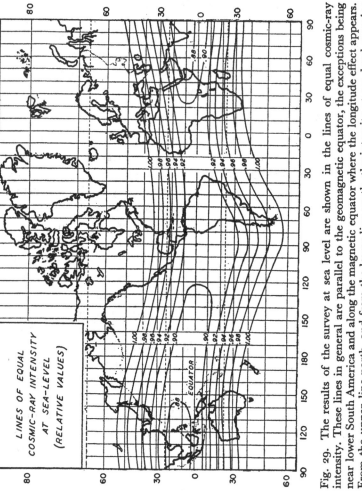

Fig. 29. The results of the survey at sea level are shown in the lines of equal cosmic-ray intensity. These lines in general are parallel to the geomagnetic equator, the exceptions being near lower South America and along the magnetic equator where the longitude effect appears. From the upper line north, and from the lower line south, the intensity at sea level remains constant save for magnetic storms and atmospheric temperature effects in winter.

eleven voyages undertaken with Neher instruments to obtain the variation of sea-level cosmic-ray intensities the world over. It was the first such chart ever published.* It is a chart which reflects the properties of the earth's magnetic field far out in space, rather than merely at the earth's surface as does a nautical chart. We shall return again to the longitude effect a little later.

4. *The east-west effect.*

The earth's magnetic field was adapted not only to the determination of whether there is a charged particle component in the incoming cosmic rays, but also, after this question had been answered affirmatively, to the answering of the further question as to what is the sign of charge that predominates in the incoming particles. Thus in the equatorial belt, for example, where the longitude effect proves that there are incoming charged particles of sufficient energy to reach the atmosphere through the blocking effect of the earth's magnetic field, if these particles are predominantly positives, then, since the magnetic lines of force above the earth are running in the direction from south to north, in accordance with the right-hand, or "motor rule", charged positive particles coming in, say vertically, must experience a force pushing them toward the east, so that they will strike the earth's atmosphere coming, not from the vertical, but from a direction inclined to the west of the vertical. Negatively charged particles, on the other hand, will strike the earth coming predominantly from a direction inclined to the east of the

* Carnegie Institution of Washington, *Year Book*, No. 34, p. 343 (13 December 1935); also *Phys. Rev.* **50**, 24 (1936).

vertical. So-called Geiger-Müller counter tubes are admirably adapted to determining the direction in which charged particle rays are coming into the earth. If two such tubes, say an inch in diameter and 6 inches long, are set up pointing north and south, one above the other and say a foot apart, mounted on a vertical rod or frame that can be tilted either toward the east or toward the west, and if the incoming charged particles at the equator are predominantly positives, more of them will pass through both counters and hence more responses will be registered when the vertical frame is tilted toward the west than when it is tilted toward the east. If the incoming charged particle rays, on the other hand, are predominantly negatively charged, there will be a larger number of rays passing through the counters when the inclination is toward the east than when it is toward the west. Both Rossi and LeMaitre and Vallarta had suggested this experiment to be carried out somewhere in or near the equatorial belt, and indeed Rossi had looked for this effect without finding it.* It was brought to light independently by T. H. Johnson† of the Bartol Foundation, and Alvarez and Compton‡ of the University of Chicago, both of whom in the spring of 1933 set up such counter tubes in the mountains near Mexico City and found when they tilted their vertical counter-tube system toward the west they got more counts than when they tilted it toward the east, thus proving that

---

* Rossi, *Phys. Rev.* **36**, 606 (1930); *Rend. Lincei*, **13**, 47 (1931) and *Nuovo Cim.* **8**, 85 (1931); also *Phys. Rev.* **45**, 212 (1934).

† Johnson, *Phys. Rev.* **43**, 834 (1933); also *ibid.* **48**, 287 (1935), for most extended article.

‡ Alvarez and Compton, *Phys. Rev.* **43**, 835 (1933).

the incoming charged particles are predominantly positive. This result was surprising in view of the fact that until the discovery of the positive electron by Anderson, positive charges had been found only in the nuclei of atoms, and in general nuclei are very much less abundant than are the negative electrons, which constitute the extra-nuclear shells of the atoms. The abundance of these negatives and their ease of detachment from atoms would seen to favour their predominance in the field-sensitive component of the cosmic ray, but in any case the observed direction of the east-west effect said unambiguously that positives of some sort predominated. Rossi,* Korff,† Neher‡ and Clay§ have all checked and extended these experiments, and they are all in essential agreement about the foregoing results. It is an experiment the interpretation of which will probably some time be of fundamental importance for our understanding of cosmic processes.

Quite irrespective of the sign of the incoming particles, however, even if they are all of one sign, no east-west effect can be found for rays that are inside Epstein's and Lemaitre's polar cap of uniform cosmic-ray intensity, which it will be remembered holds for rays of a given energy down to a particular latitude, this latitude moving farther and farther south for increasing values of the energy of the incoming particles. This means that the east-west effect at sea level in the Americas should only be observed south of about mag lat 41°. It is, in fact, a maximum at

* Rossi, *Phys. Rev.* **36**, 606 (1930); *Rend. Lincei,* **13**, 47 (1931) and *Nuovo Cim.* **8**, 85 (1931); also *Phys. Rev.* **45**, 212 (1934).

† Korff, *Phys. Rev.* **46**, 74 (1934).   ‡ Neher, not yet published.

§ Clay, *Physica,* **1**, 363 (1933).

the magnetic equator, as T. H. Johnson has experimentally proved. The earth's magnetic field of course acts upon charged particles inside the polar cap just as much as outside it, but the uniform intensity inside that cap means that this field here causes as many rays to be bent into a given angle of incidence on the earth's surface as out of it. Table 3* shows just how the east-west effect is brought to light and how measurements upon it are made. The percentage asymmetry A is taken simply as the difference

TABLE 3. East and west counts taken by T. H. Johnson at Cerro de Pasco, Peru, altitude 4300 metres on the magnetic equator

Zenith angle Z	Total counts west C_w	Total counts east C_e	Asymmetry A	Probable error of A
15°	59,295	55,049	0·084	0·0042
30°	39,601	35,418	0·125	0·0054
45°	11,024	9,764	0·139	0·0085
60°	6,760	6,028	0·174	0·015

of the east-west counts divided by half their sum. At the equator this asymmetry A does not appear to be appreciably different at sea level and at the elevation of Cerro de Pasco, namely, 4300 metres, nor is there any reason to expect it to be. However, just as the polar cap of uniform cosmic-ray intensity was found in our airplane survey to extend farther south at sea level than at high altitudes so, *in these intermediate latitudes*, the east-west effect should increase with altitude, as Johnson found it to do in his observations in Mexico. We shall have more to say about the east-west effect a little later.

* See Johnson, *Phys. Rev.* **48**, 290 (1935).

5. *A new latitude survey extending practically to the top of the atmosphere.*

The difficulty of extending latitude measurements on cosmic rays higher than to altitudes of about 30,000 feet, about the limit which we were able to reach with airplanes, lay in this fact that in order to obtain the desired accuracy in high-altitude intensity measurement it was necessary to recharge our Neher self-recording instruments every 4 or 5 minutes during a flight, and this we had accomplished in airplanes by taking up a 350 volt battery which was used as the charging source. In order to work at the higher altitudes attainable with sounding balloons it was necessary to develop the technique of repeatedly charging during flight the Neher electroscope from a condenser instead of from a battery, since the latter had a prohibitive weight. After a couple of years of experimenting, with the aid of Dr S. K. Haines, the problem was solved of producing a very light charging condenser of 10,000 cm capacity, which during the conditions existing in a flight lost less than a tenth of a per cent of its charge per hour.

With these new Neher instruments weighing, with all the recharging and recording mechanisms for electroscopes, barometers and thermometers, only about 1400 grams, in August 1936* we made from Fort Sam Houston, San Antonio, Texas (mag lat 38° 30′ N), with the aid of Colonel Prosser, Lieutenant Matthews, and the U.S. Signal Corps unit stationed there, successful flights which gave accurate records of cosmic-ray intensities up to 12·9 mm of mercury, or 98·3 % of the way to the top of

* *Phys. Rev.* **50**, 992 (1936).

Fig. 30. Launching electroscope for stratosphere flights. In this ascent only four balloons were used. In reaching our highest altitudes (9·9 mm of mercury) we have regularly used ten balloons.

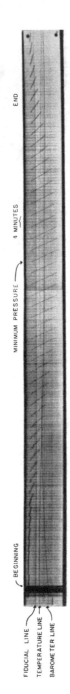

FIDUCIAL LINE
TEMPERATURE LINE
BAROMETER LINE

BEGINNING

MINIMUM PRESSURE

4 MINUTES

END

Fig. 31. A typical record taken at Omaha, Nebraska, in the summer of 1937. The slope of the discharge line is a direct measure of the rate of discharge, i.e. the ionization. The electroscope is automatically charged up every four minutes from a non-leaking condenser of 10,000 cm capacity. The record shows how accurately the cosmic-ray ionization at any altitude (barometer line) can be obtained.

the atmosphere.* We then at once shipped the equipment with Dr Neher to Madras, India (mag lat 4° N), and with very fine co-operation from the Indian Meteorological Service repeated there, very close to the magnetic equator, the type of measurement already made in San Antonio. The results of these measurements enabled Bowen, Millikan and Neher to answer at once one of the most disturbing questions in the whole field of cosmic rays—one which had given us all a headache for at least five years. The history and significance of the question is as follows:

From the time of the first quantitative proof brought forward by Chao† in 1930 that a new kind of absorption of γ-ray photons by lead begins to set in at an incident energy of about 2 million e-volts, the theorists had been trying to work out the possible laws of such nuclear absorption, for prior to this the laws of extra-nuclear absorption, represented by the Klein-Nishina formula, had been supposed to be adequate to take care of the whole absorption process. Oppenheimer and Plesset‡ in 1933 seized on the newly discovered fact of positive-negative pair formation through the impact of photons on the nuclei of atoms and made it the basis of a theory of the nuclear absorption of photons, and Lauritsen and his group at the California Institute of Technology had shown that this theory was competent to describe the facts of the

* Pfotzer's remarkable flight, made with triple vertical counters on 2 November 1935, reached to 10 mm of Hg (see *Zeit. f. Phys.* **102**, 23 and 41 (1936)). Also in the summer of 1937 we reached altitudes at which the pressure was 9·9 mm of Hg, or 98·8 of the way to the top of the atmosphere (see *Phys. Rev.* **53**, 855 (1938)).

† Chao, *Phys. Rev.* **36**, 519 (1930).

‡ Oppenheimer and Plesset, *Phys. Rev.* **44**, 53 (1933).

absorption of the photons in the whole energy range from 2 million up to 12 million e-volts which they were then opening up to exploration. But the trouble with the theory as applied to cosmic rays was that it did not permit even billion volt rays, whether consisting of photons or electrons, to penetrate through more than a metre or two of water at the most. Bethe and Heitler* worked out the theory in more detail, especially as it applied to the absorption of high-energy electrons. The fundamental assumption underlying this theory, now generally known as the Bethe-Heitler theory of the absorption of electrons, has already been given (see p. 35). The essential point is that electrons, in trying to plunge through matter, in view of their extreme lightness, must be very sharply accelerated or decelerated as they plunge into or past a relatively heavy object like an atomic nucleus, and this act of deceleration, or "bremsung", must transform the incident energy into an electromagnetic pulse, a "scattered X-ray photon", which by quickly encountering another nucleus produces an electron pair (+ and −), each member of which pair repeats the original process, etc., so that the energy is very rapidly frittered away by the continual repetition of these "bremsstrahlung"—pair formation—acts.

Now this Oppenheimer-Bethe-Heitler theory of the absorption of both photons and electrons worked quantitatively everywhere when tested in the range of energies of a few million e-volts.† It required the loss of energy of an

---

* Bethe and Heitler, *Proc. Roy. Soc.* **146**, 83–112 (August 1934).

† Crane and Lauritsen, *International Conference on Physics*, I, Nuclear Physics, pp. 130–43 (London, 1934).

electron or photon in going through a given thickness of matter to be proportional to the incident energy—in other words, it required the absorption *coefficient* to be constant no matter what was the value of the incident energy, and it required the absorption in going through a space containing a given number of atoms of different atomic number to be proportional to Z^2, where Z means atomic number. Both of these conditions appeared to be satisfied when the rays experimented upon were of low energy, i.e. from 2 million up to perhaps 300 million e-volts.* But it obviously did not hold for high-energy cosmic rays. Why not? Because it did not permit *the cosmic-ray photons or electrons to penetrate through more than a few centimetres of lead or a couple of metres of water at the most*, despite the fact that ever since Millikan and Cameron's original experiments in snow-fed lakes it had been proved beyond question that cosmic rays *do* penetrate not merely through 1 atmosphere—10 metres of water—but through not less than 10 atmospheres, i.e. to depths in lakes of at least 300 feet (100 metres) of water. Again, while the above theory did not permit electrons of any energy, however large, to pass through more than a couple of centimetres of lead, Wilson cloud chambers showed tracks that were apparently indistinguishable from electron tracks penetrating without appreciable change in direction through 15 or 20 cm of lead. Further, I have already indicated that cosmic rays as found at sea level seem to follow a law of absorption in different substances which is roughly proportional to density. This is nearly the same as a Z law, instead of a Z^2 law, as required by the theory. Only in measuring

* Anderson and Neddermeyer, *Phys. Rev.* **50**, 256 (1936).

shower phenomena had a Z^2 law come to light through the work of Rossi and others, but these phenomena seemed to be produced in greatest abundance by a soft or non-penetrating component of the cosmic rays, a component that reached its maximum of shower-producing effectiveness at a thickness of lead of about 1·6 cm, so that it seemed that we might be justified in assuming that the Bethe-Heitler theory held for low-energy rays, but that it did not hold for any high-energy rays.

This was the simplest way out of the difficulty, and the way I myself took.* Indeed, as late as the autumn of 1936 I called attention to the above-mentioned "range" phenomena connected with the cosmic-ray shelves, one near Pasadena at sea level and another near Spokane at an altitude of 21,000 feet, and saw in them the proofs of the failure of the Bethe-Heitler law at energies of a billion volts and more. Also, all of the theorists, Bethe, Heitler and Oppenheimer, expressed the belief that their law could not hold for very high energies.

The new high-altitude experiments of Bowen, Millikan and Neher, however, have shown that we were all wrong and that the breakdown of the Bethe-Heitler law at high energies is not the way out of the foregoing apparent cosmic-ray contradictions.

The experimental justification of the foregoing statement is found in the graphs shown in Figs. 32, 33 and 34, which give at various latitudes the full curves, up to within a per cent or two of the top of the atmosphere, of the relationship holding between the number of cosmic-

* Millikan, Carnegie Institution of Washington, *Year Book*, No. 35, p. 349 (11 December 1936).

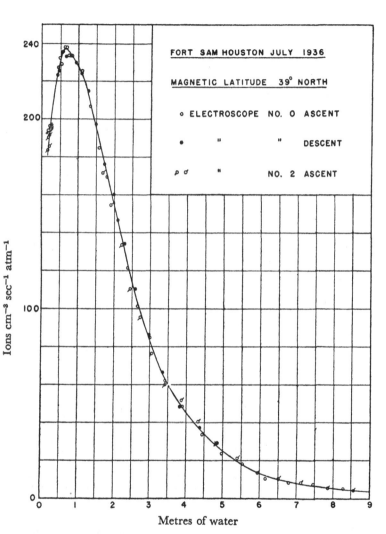

Fig. 32. Altitude-ionization curve for both flights, reduced to ions per cc at atmospheric pressure. The pressures are in metres of water below the top of the atmosphere (10 m = 1 atmos).

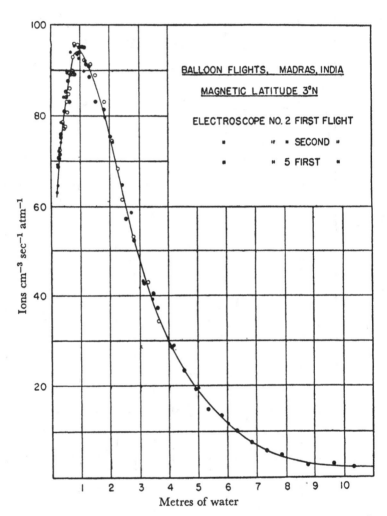

Fig. 33. Ionization as a function of depth, in equivalent metres of water, below the top of the atmosphere at Madras, India, mag lat 3° N.

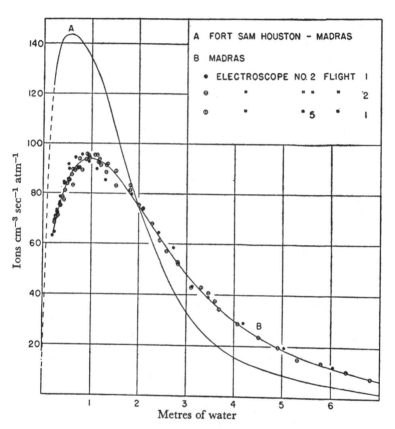

Fig. 34. Curve A shows the ionization at all depths down to 7 m of water due to the electrons entering the atmosphere of energies between 6·7 billion e-volts and 17 billion e-volts. This curve is the difference between the curves of Fig. 32 and Fig. 33. Curve B is the same as that shown in Fig. 33 but drawn to the same scale as curve A. The areas (or incoming energies) under the two curves are nearly the same.

ray ions formed per cc per sec within our pressure electroscope, reduced to 1 atmosphere of air, and the depth beneath the top measured in equivalent metres of water, at which this electroscope is placed. The following striking and, for the most part, new facts stand out at once from an inspection of these curves.

1. Cosmic rays, whatever their nature, are so rapidly absorbed as a whole in the outer layers of the atmosphere that even in the equatorial belt (Fig. 33), where the effect of the earth's magnetic field upon them is a maximum, they get into equilibrium with their secondaries and produce their maximum ionization before they have penetrated through the first tenth of the atmosphere.*

2. From that maximum point on, in all cases they fall off exceedingly rapidly in intensity following an exponential equation, their law of absorption being like that of X-rays and not like that of particles that exhibit range phenomena such as low-energy β-rays, proton rays, or α-rays. The apparent *range phenomena* discussed in preceding pages appear indeed at sea level and at other

* This effect was suggested as a possibility by Millikan and Cameron in 1927 (*Nature*, **121**, 20 (1927)) in their report made at the Leeds meeting of the British Association on their first voyage (1926) made from Los Angeles to Peru to look for the effect of the earth's magnetic field on incoming electrons. The words then used were: "If the northern hemisphere and the southern hemisphere curves (of ionization with altitude) coincided, it would go a long way toward eliminating the possibility that the rays are generated by the incidence of high-speed beta rays on the very outer layers of the atmosphere.... For such beta rays would be expected to be influenced by the earth's magnetic field so as to generate stronger radiation over the poles than over the equator." This is precisely what the present experiments show to be the case for the whole field-sensitive portion of the cosmic rays.

relatively low altitudes, but they are not at all characteristic of the absorption of cosmic rays in *the upper layers of the atmosphere*.

3. The depth beneath the top of the atmosphere at which the maximum ionization is attained, *always less than a tenth of an atmosphere*, changes but slightly in going from San Antonio (Fig. 32), where no electrons of energy less than 6.8×10^9 e-volts can get vertically through the blocking effect of the earth's magnetic field, to Madras (Fig. 33), where no electrons of energy less than 17×10^9 e-volts can similarly get through.

4. The difference between the San Antonio and the Madras curves makes possible for the first time the determination of the complete curve of ionization produced in the atmosphere by incoming charged particles contained within a sharply limited band of energies having a weighed mean value of 10×10^9 e-volts, and it is this curve (Fig. 34, curve A) in particular, although it will be seen to be very similar in character to the other full radiation curves (Figs. 32 and 33), that throws so much new light on cosmic-ray absorption and ionization, for we are here completely unable to do what many of us had always done before, namely, call on less and less penetrating photonic components of the incoming radiation for explaining the rapid rise in ionization with increasing altitude. The "new use of the earth's magnetic field" is *in this instance, then, that of supplying us with a definitely known band of charged particles of enormous but well-determined energies, and enabling us to see without any uncertainty just how charged particles of this mean energy of 10 billion e-volts are absorbed by the nuclei of the molecules of air.* The curve A of Fig. 34 gives us at once this

law and enables us to compare this *actual* rate of absorption with that required by the Bethe-Heitler theory as extended by Carlson and Oppenheimer (Fig. 35).

5. Down to a depth of a third of an atmosphere from the top (3·3 metres of water) this curve is in reasonably good agreement, as shown by Fig. 35, with the Bethe-Heitler theory of nuclear absorption as recently extended by Carlson and Oppenheimer, as well as by Bhabha and Heitler.*

6. The exceedingly rapid absorption of this latitude-sensitive radiation, with an absorption coefficient which is nearly constant and independent of incident energy, qualitatively justifies the "shower theory" of the foregoing authors as the main cause of the ionization of the atmosphere produced by incoming electrons even of this huge energy.

7. The latitude-sensitive part of the cosmic-ray ionization (Fig. 35) found in the *lower* part of the atmosphere is considerably more penetrating than is predicted by the foregoing extended Bethe-Heitler theory of electron absorption; nevertheless, while at a distance of one-twentieth of an atmosphere from the top, these 10×10^9 e-volt field-sensitive rays are producing 144 ions per cc per sec (curve A, Fig. 34), at *sea level* their total ionizing influence has fallen to but 0·3 ion per cc per sec, that is, to less than 1/500th of their value near the top of the atmosphere.

8. The two foregoing results in 7 show that the process of nuclear absorption of electrons is more complicated and involves the production of more penetrating secondaries

* Bhabha and Heitler, *Proc. Roy. Soc.* **159**, 432 (1937). See also Bhabha, *ibid.* **164**, 257 (1938).

than is pictured in the simple physical assumptions underlying the Bethe-Heitler theory, but, at the same time, that the whole progeny of secondaries, whatever their nature, has been reduced almost to zero by the time sea level has been reached, not more than about one-tenth of the sea-level ionization being accounted for by field-sensitive rays at all.

9. The latitude-sensitive part of the cosmic-ray ionization found in the lower atmosphere is practically all due to the *secondary* effects of varied nature resulting from the absorption of the incoming electrons in the upper tenth of the atmosphere.

10. The *apparent* absorption coefficient, namely, 0·54 per metre of water, of the actual curve (A, Fig. 34) representing the whole progeny of secondary influences resulting down to sea level from the absorption of incoming electrons in the very top layers of the atmosphere (for this one coefficient fits fairly well the whole curve from sea level, or 10 metres, up to 2·5 metres of water) is approximately the same as that found by Johnson and by Neher for the east-west effect, thus proving that the particles causing the latitude and the east-west effect are of the same type. Both absorption coefficients are such as to suggest that these particles are electrons (predominantly positive), not protons.

The experiments and conclusions thus briefly summarized leave no sort of uncertainty as to the wrongness of my former assumption that the observed very great penetrating power of charged particles in cosmic-ray phenomena is to be explained by the breakdown at energies of the order of a billion or more of volts of a law like the Bethe-Heitler law which requires the very rapid absorption of electrons. As the Wilson cloud-chamber experiments directly demonstrate, there are indeed penetrating particles present in sea-level cosmic rays to which no law of rapid absorption like the Bethe-Heitler can possibly apply. To that extent the cosmic rays do reveal a breakdown of that theory, but since we have proved by these experiments that even 10 billion volt *electrons* are in fact very rapidly absorbed in air, essentially as the fundamental assumptions underlying Bethe-Heitler require, or perhaps even more rapidly, there is no alternative left but to conclude that these highly penetrating rays observed in cloud chambers and required by under-water effects *are not electrons*, for they show a penetration that electrons even of their energies do not possess. But what, then, are they?

Since it is *mass* or *inertia* (inability to be suddenly "bremsed", i.e. inability to produce "radiative collisions") that constitutes the essential difference between the absorbability of electrons and heavy particles, we have no alternative but to classify these highly penetrating charged particles observed in Wilson cloud chambers at sea level, and showing great penetrating power below sea level, and to a lesser degree above it, as *heavy particles of some kind*. Auger and many other Europeans have for years looked upon them as *protons*, but since these penetrating particles, as shown in Lecture II, are about equally positive and negative in sign, if they were protons then these cosmic-ray experiments would have brought to light another new particle, namely, a *negative proton*. There is, however, the best of evidence that some of them at least, I think almost all of them, cannot be as heavy as protons, in that they seem to be able to transfer their energy to extra-nuclear electrons more easily than a body as heavy as a proton should be able to do. Therefore, in order to leave their precise nature to a degree unsettled, Anderson and Neddermeyer at first simply called them X-particles, their sole new characteristic being that they were heavy enough not to be "bremsed", i.e. not to experience radiative collisions and thus have their energy transformed into photons. This is all that is left to call upon to make a particle penetrating if we are obliged, as we seem to be by the foregoing experiments, to give up the notion that high energy alone can make an electron penetrating. It is true that the tracks of these penetrating particles are thus far indistinguishable from electron tracks and their charge is certainly the same as that of the electron, and they are like the electron in that their charge may be either positive or negative in sign. There may be those who will still wish to believe that they are electrons, but if so they are very penetrating electrons and the foregoing experiments seem definitely to deny high penetrability to electrons up to the huge energies of 10 billion e-volts. The cosmic rays have been teaching us some strange new things about nature that we had not even dreamed of before.

6. *The origin of the penetrating particles.*

But however incomplete the evidence may yet be as to whether these penetrating particles found in cosmic rays near sea level and below are in part positive and negative protons, or whether they are all positive and negative particles of intermediate mass between protons and electrons, i.e. mesotrons, there is at any rate considerable evidence as to where most of them come from. In a word, the evidence is that at least the great majority of those that we find in the lower part of the atmosphere and below sea level do not come in from outside at all, as all of us originally assumed them to do, but that rather they are for the most part *produced* as secondary penetrating particles in our atmosphere by some kind of nuclear collisions that other rays, presumably photons, make with the nuclei of the atoms constituting the atmosphere. The evidence for the correctness of this view comes from a number of directions, as follows:

First, the most simple and direct evidence comes from the comparison of cloud-chamber tracks with experiments on the east-west effect. These last show, as already indicated, that the incoming charged particles are predominantly positives, yet according to all the cloud-chamber observers* the penetrating particles observed at sea level are nearly equally divided between positives and negatives. For example, Blackett makes the percentage of positives 53 ± 2. This *equating* by passage through the atmosphere of positives and negatives when the incoming rays are strongly of the positive sign seems to mean that the chance that the incoming rays will produce in the atmosphere a penetrating negative is about the same as the chance of producing a penetrating positive.

* Anderson, *Phys. Rev.* **44**, 406 (1933); Anderson and Neddermeyer, *International Conference on Physics*, I, Nuclear Physics, p. 173 (London, 1934); Kunz, *Zeit. f. Phys.* **80**, 559 (1933); Blackett and Brode, *Proc. Roy. Soc.* **154**, 573 (1936); LePrince-Ringuet et Jean Crussard, *C.R.* **204**, 112 (1937).

The second sort of evidence is provided by the upper curve of Fig. 34. This represents all the ionizing influences, primary and secondary, simple and complicated, which all the incoming charged particles between the energies 6.7×10^9 e-volts and 17×10^9 e-volts, *whether they be protons or electrons*, are able to produce in our atmosphere, for protons and electrons are influenced in essentially the same way by the earth's magnetic field. It will be seen from the curve that, as already indicated in the foregoing summary, at a depth of 0·5 metre of water beneath the top, i.e. at one-twentieth of the way through the atmosphere, 144 ions per cc per sec are being produced by these incoming charged-particle rays, while by the time sea level has been reached only 0·3 of one ion per cc per sec, or 1/500th the maximum number, is being produced. In other words, nearly all of the effects of all the incoming particles, protons and electrons and mesotrons combined, between the energy limits 6.7×10^9 e-volts and 17×10^9 e-volts, have been wiped out in going through the atmosphere. Even if the whole of the effect left at sea level is due to the incoming penetrating particles, and no influence whatever of the electrons has reached down this far, the number of entering penetrating particles is then necessarily very minute and their atmospheric ionizing influence is utterly negligible in comparison with that of the incoming electrons. In other words, *there cannot in any case be any appreciable number of penetrating particles of any kind coming in with the electrons of range from 7 billion to 17 billion e-volts.* The total vertical resistance of the atmosphere to such penetrating rays computed from the ionization along the ion track and the number of δ-rays that they can produce is reliably estimated at about $2\frac{1}{2}$ billion e-volts, so that if they enter in appreciable numbers in the energy range 7 billion to 17 billion e-volts their ionizing influence would be large instead of negligibly small.

But the curve itself bears internal evidence that the incoming *electrons* themselves possess the power of producing, by some direct or indirect process as their influence travels through the atmosphere, particles more penetrating than themselves. On the logarithmic scale shown in Fig. 35 the full curve shows the variation of ionization with depth beneath the top of the atmosphere as it is predicted in our electroscope by the two fundamental mechanisms underlying the Bethe-Heitler theory, namely, impulse-radiation (bremsstrahlung) and pair formation. The fact that the experimental points reach their maximum nearer the top of the atmosphere than the theoretical curve, indicates that at the top of the atmosphere the actual absorption of the energy of the incoming electrons and the consequent multiplication of ionization take place *even more rapidly* than the Bethe-Heitler theory predicts. This means that there are other mechanisms of electron absorption and consequent ion formation besides those postulated in the theory. Some of these are listed below. But the lower half of the curve in Fig. 35 shows that there, on the contrary, the ionization is *not dying out nearly as rapidly with depth as the Bethe-Heitler mechanism demands*. In other words, the radiation is becoming more and more penetrating as it sinks farther into the atmosphere. *This means merely that there is some mechanism by which more penetrating carriers of the energy downward are being produced in the atmosphere by the incoming electrons.*

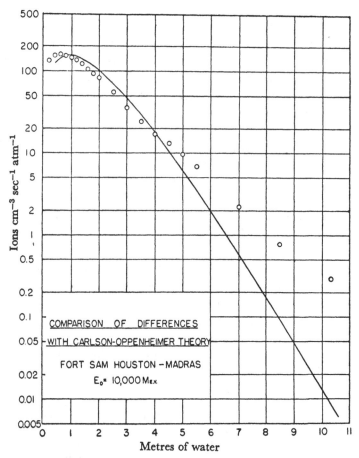

Fig. 35. Comparison of Carlson-Oppenheimer theory of ionization at different depths (full line) due to incoming electrons of mean energies 10,000,000,000 e-volts with directly observed ionizations (small circles). The scale is logarithmic. The fit is fairly good at the top, but from 5 m down to sea level (10 m) the ionization falls off much less rapidly—i.e., the rays are much more penetrating—than the theory permits.

It will be seen from the foregoing that the nuclear absorption of incoming electron energy must be quite a complex phenomenon. Indeed, there is already evidence for the existence of at least four different mechanisms of such absorption, which may be listed as follows:

1. The first is impulse radiation and pair formation. This seems to be the great, predominant influence in the higher parts of the atmosphere and the chief agent in general of nuclear absorption.

2. The second is the type of absorption illustrated by Fig. 15, that is, by Hoffmann bursts. The evidence brought forward by Carmichael* seems to be strong that it may happen, though very rarely, that the whole energy of 2 or 3 billion volt photons may be transformed in a Hoffmann burst by one nuclear encounter into the energy of several thousand low-energy electron-ray tracks (+ and −). This means that the nuclear absorption of electron energy should be *more* rapid than the Bethe-Heitler law predicts.

3. The third type of absorption is that first brought to light in Anderson-Neddermeyer photographs, which

* Carmichael, *Proc. Roy. Soc.* **154**, 123–245 (March 1936). The essence of Carmichael's conclusion is that "some process must exist by which a single high energy cosmic ray is replaced suddenly by thousands of separate ionizing particles". The largest burst he observes in 1500 hours with a vessel of 175 litres capacity is no larger than we observe with a pressure electroscope of 1·6 litres capacity. The implication is that practically the whole energy of the burst is released in the ionization chamber itself. Carmichael never finds this amounting to more than a very few billion e-volts. In other words, once in 100 hours a collision may occur in the wall so close to the inner surface as to release most of the energy of the incoming ray within the chamber itself.

showed a nucleus exploding with the emission of a relatively low-energy proton (see Fig. 18).

4. The fourth type of absorption is the hypothetical production or creation—directly or indirectly, presumably through the encounter of photons with nuclei—of the penetrating particles which are chiefly, if not wholly, responsible for the transfer of the incoming electron energy to sea level and below. These penetrating particles have the power, presumably through new nuclear impacts, both of producing new penetrating particles (+ or −) and also of transferring their energy to new photons and electrons and thus carrying showers down to very much greater depths than the Bethe-Heitler theory would otherwise permit. We need no better evidence for the existence of some mechanism that can transfer shower effects deep down into the atmosphere where the simple Bethe-Heitler theory denies them the possibility of reaching than is provided by Figs. 6 and 7, which, while taken at sea level, show showers consisting of many electron tracks (+ and −) which are produced by nuclear encounters made in the lead by 1½ billion and 2 billion volt photons and electrons. Such showers are found by Weischedel*, not abundantly but occasionally, at great depths under water. We have, then, direct, unambiguous evidence that "the penetrating particles" have the power of carrying down the energy required for these shower effects and releasing it by nuclear encounters of some sort. So far as the above-mentioned photographs alone are concerned, they merely show that some process exists by virtue of which cosmic

* Weischedel, *Zeit. f. Phys.* **101,** 744 (1936).

energy can be carried down and released as a photon-
made shower or an electron-stimulated shower at depths
to which the Bethe-Heitler theory does not permit photons
or electrons to penetrate, especially photons and electrons
of billions of volts of energy. Our present assumption is
that the penetrating particles which may be produced
occasionally farther up by the impact of a photon on a
nucleus may act as such carriers or links in the chain of
transmission of energy downward.

To summarize, then, the argument from the character
of the curve shown in curve A, Fig. 34. This curve is
satisfied by one single apparent absorption coefficient,
namely, $\mu = 0.54$ per metre of water, from sea level, or
10 metres of water up to a depth beneath the top of about
2.5 metres (see Fig. 35), and its character is not such as to
be consistent with the view that any appreciable number
of protons or penetrating particles of any kind accompany
the incoming electrons of energies between 6.7 billion
and 17 billion e-volts, but it does seem to demand some
mechanism for the production of penetrating particles
directly or indirectly by the electrons themselves. Certainly
in all its significant features *this whole latitude curve is due
to incoming electrons and their progeny of secondaries of one sort
or another*.

That similarly the electrons of somewhat higher energy
which penetrate into the equatorial belt and are responsible
there for the east-west and longitude effects are not ac-
companied by any appreciable number of protons or
other highly penetrating particles is shown by the measurements that T. H. Johnson made at the magnetic equator
on the absorption of the atmosphere for the positive rays

responsible for the east-west effect. The main reason why the counting rate of a system of double or treble coincidence counters arranged in a vertical line falls off so enormously in tilting it from the zenith down to the horizon lies in the varying amount of atmosphere through which the rays must pass at various zenith angles in coming to the counter system, and the differences between this absorption east and west enabled Johnson to compute the absorption of the atmosphere for the incoming positives. He thus got for various latitudes in the equatorial zone values varying between 0·33 and 0·53 per metre of water. But this is so close to the single apparent absorption coefficient which holds for curve A, Fig. 34, namely, 0·54 per metre of water from bottom to well up toward the top, and which has just been shown to be due to incoming electrons as to leave little doubt that the incoming particles with which Johnson was dealing were also positive electrons and not heavier particles. This is the third bit of evidence that the penetrating particles observed at sea level do not enter in appreciable numbers from outside but are in the main secondaries created in the atmosphere. At sea level where Johnson worked in the equatorial belt he should have got the absorption coefficient for his positives corresponding to protons if these were the particles which came through the atmosphere and produced his positive excess. But according to the Bethe-Heitler theory this coefficient is of the order of a millionth that for electrons, for it is inversely proportional to the square of the mass of the particle, so that no appreciable number of incoming protons seem to be admissible in Johnson's experiments at the equator. He was clearly dealing with the same kind of incoming particles as are responsible for the latitude effect, since in measuring their total summed effects, primary and secondary, from the top of the atmosphere down to sea level, he finds approximately the same apparent absorption coefficient as we find for the right-hand side of curve A, Fig. 34.

It is to the penetrating rays created by some indirect process by the bombardment of the nuclei of the upper air by electrons that we must look for our interpretation of the existence of a sea-level cosmic-ray shelf in latitude 41°, for the apparent change as sea level is approached from something like a Z^2 law of absorption to a Z absorption law, and for the understanding of all the apparent range properties, cosmic-ray shelves, etc., which come into play so prominently near sea level and below it, as shown in Figs. 34 and 35.

The foregoing unquestionable evidence that practically all of the *field-sensitive cosmic rays* enter the atmosphere not as penetrating (heavy) particles of any kind, but solely as electrons, is then found in the shape and character of the upper curve of Fig. 34. But the evidence that is found in the shape and character of the curve of Fig. 33, and this represents the effects of all the *non-field-sensitive* cosmic rays, is nearly as convincing that all of these rays, too, are of a highly absorbable type. This means that though they are not necessarily all electrons, for they are not measurably deflected by the earth's magnetic field, they must be either a mixture of photons and high-energy electrons or else of high-energy electrons alone.*

The net result, then, of all the considerations advanced

* This evidence is presented at length in *Phys. Rev.* **53**, 217 (1938).

in this section is contained in the statement that the evidence is good that *practically all of the cosmic-ray effects observed in the lower part of the atmosphere are secondary effects—splashes of various kinds—produced in the upper layers of the atmosphere by the inflow from outside of electrons (+ and −) or of electrons and photons combined, which, no matter what their energy, cannot themselves penetrate through the upper layers because of the powerful barrier set by the laws of nuclear absorption.*

Our experiments have thus far yielded no crucial evidence as to the relative roles played by electrons and photons in producing the ionization due to the non-field-sensitive half as measured by energies (see below) of the incoming cosmic rays. A certain part, however, of the non-field-sensitive rays must in any case be electrons in order to account for the equatorial east-west effect.

7. *The distribution of energies among incoming cosmic rays.*

The measurement in electroscopes of the ionization produced at all altitudes clear to the top of the atmosphere makes it possible to determine the distribution of incoming cosmic-ray energies—a forward step of much interest and significance. The procedure is as follows: Since the right side of curve A of Fig. 34 falls off exponentially and has at sea level (10 metres of water) a value of only 0·3 ion, it is clear that the area below sea level (or beyond 10 metres) between the curve and the X axis is altogether negligible in comparison with the corresponding area above sea level. The directly observed curve is seen (A, Fig. 34) to extend nearly to the top and the dotted line extends it with very little uncertainty* to the very top. Hence this

* See *Phys. Rev.* **53**, 217 (1938) for more complete discussion.

area underneath curve A, when that curve is extended down to sea level, is simply the integral of all the ions that all the incoming electrons of energies between 6·7 billion e-volts and 17 billion e-volts are able to produce per cc per sec within our electroscope as it goes from sea level to the top. This actually comes out $2·8 \times 10^7$ ions. Since all the energy of these incoming electrons is expended in producing ions, we may then take this area, or this number of ions, as a measure of the total energy brought into the earth per sec per sq cm of the earth's surface by all of the incoming electrons that are included within the foregoing energy range. If we wish to reduce this to e-volts we can do so by multiplying by 32, the number of e-volts required to produce an ion pair in air. We obtain thus 9×10^8 e-volts. Again, if we wish to find how many electrons within this range of energies hit each sq cm of the surface of the earth (upper atmosphere) per sec, we have only to divide by the mean energy included between $6·7 \times 10^9$ and 17×10^9, namely, about 10×10^9, and we thus find quite definitely that *one such 10 billion volt electron enters each sq cm of the surface of the upper atmosphere every 11 seconds*. We are thus obtaining quite precise and dependable information about the number of cosmic rays of certain definite energies that shoot through space.

We can analyse in much the same way the curve shown in Fig. 33 and repeated in the lower curve of Fig. 34. However, since this curve taken at Madras near the magnetic equator is due wholly to non-field-sensitive rays, we have no means of telling how much of the inflowing cosmic-ray energy is here due to photons and how much to electrons of energy higher than 17 billion e-volts—too high to be measurably influenced by the earth's magnetic field. But we can know that the total area beneath this curve represents the whole of the incoming non-field-sensitive cosmic-ray energy. And when we measure up this area we find it roughly the same as that of curve A, Fig. 34. More accurately, taking account of sub-sea-level ionization, it is 8% larger than the area of the latter curve. We know, then, altogether definitely that *the total cosmic-ray energy brought in by electrons of energy above 17 billion e-volts plus all that brought in by photons of all energies is about the same as the energy brought in by the electrons alone of energies between 6·7 billion e-volts and 17 billion e-volts.*

The next experimental step that was clearly indicated was to take depth ionization curves similar to those shown in Figs. 32 and 33 at a series of latitudes extending as far toward the north magnetic pole as possible so as to find the complete distribution of energies of all the incoming electrons. In August and early September 1937 we accordingly made eleven new flights in which we reached a minimum pressure of 9·9 mm of mercury, equivalent to 98·8% of the way to the top of the atmosphere. Three of these successful flights were made at Saskatoon, Canada (60° N mag) and six at Omaha, U.S.A. (51° N mag). The duplicability of the readings obtained on different flights, and hence the dependability of the final curves, can be estimated from a comparison of the figures in each horizontal row in Tables 4 and 5. Also, Fig. 31 shows a typical developed film obtained in one of these flights with a Neher electroscope. The latter was automatically charged up every 4 minutes from a condenser which lost not over 0·5% of its charge per hour. The total free weight lifted by the ten balloons arranged in tandem was about 1900 grams. The instrument alone weighs 1400 grams. The film shows the record obtained during a flight of 3 hours 20 minutes duration. The altitude at which at the end of about 2 hours the descent began because of the

TABLE 4. Comparison of flights at Saskatoon, Canada

Metres of water	Film No.			Average
	1k20	4k20	9k20	
0·2	355	—	—	355
0·3	363	356	—	359
0·4	362	360	—	361
0·5	357	358	—	357·5
0·6	350	349	341	347
0·8	333	330	323	329
1·0	311	307	305	308
1·2	279	279	282	280
1·4	250	252	254	252
1·6	224	225	227	225
1·8	199	203	203	202
2·0	177	181	181	180
2·25	—	155	156	156
2·50	—	—	134	134
2·75	—	—	115	115
3·0	—	—	97	97
3·5	—	—	68	68
4·0	56	—	47	51
4·5	40	—	32	36
5·0	28	—	22	25
5·5	20	—	16	18
6·0	14	—	11	13
7·0	10	—	—	10

bursting of balloons can be seen on the film. The combined results of all these flights at different latitudes is found in Fig. 36. This shows on the same scale the four depth-ionization curves reduced from an electroscope filling of argon at a pressure of 2 atmospheres to air at 20° C, 76 cm of Hg, taken in the four different magnetic latitudes, namely, those of Madras (3° N), San Antonio (38° N), Omaha (51° N), and Saskatoon (60° N).

The entirely new and quite unexpected result that came to light from this comparison was the relatively slight difference between the Omaha and Saskatoon curves.

TABLE 5. Comparison of flights at Omaha, Nebraska

Metres of water	Film No.						Average
	2k20	2k21	4k21	6k21	6k22	9k21	
0·2	—	—	—	—	328	—	328
0·25	—	—	—	—	333·5	—	333·5
0·3	—	—	—	—	337	—	337
0·4	—	—	323	335	340	—	333
0·5	330	338	322	338	337	—	333
0·6	321	330	317	335	330	—	326
0·8	304	309	304	322	314	304	310
1·0	285	286	287	301	294	290*	290
1·2	264	261	267	276	267	—	267
1·4	242	236	242	249	241	—	242
1·6	219	213	215	223	217	—	217
1·8	196	194	191	199	193	—	195
2·0	175	175	168	177	171	—	173
2·25	150	152	143	152	146	—	150
2·5	128	131	122	131	123	—	128
2·75	109	113	—	113	104	—	110
3·0	92	96	—	98	89	—	94
3·5	68	72	—	70	66	—	69
4·0	50	—	—	49	48	—	49
4·5	37	—	—	33	35	—	35
5·0	28	—	—	25	26	—	26
5·5	21	—	—	19	—	—	20
6·0	16	—	—	—	—	—	16
7·0	11	—	—	—	—	—	11

* Mean of sixteen 4 minute discharges.

This means that although the blocking effect of the earth's magnetic field is reduced in going from Omaha to Saskatoon from about 2·9 billion volts to 1·4 billion volts, yet there is very little new electronic energy that comes into the atmosphere in this energy range.

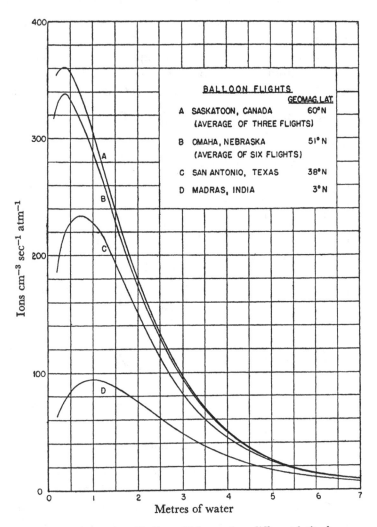

Fig. 36. Results of balloon flights at four different latitudes.

We have just recently, namely, about 1 July 1938, taken similar depth-ionization curves at Bismarck, North Dakota (56° N magnet lat), and found the curve there quite as high as at Saskatoon—apparently a bit higher. Accordingly the conclusion seems justified that although cosmic-ray electrons of energies higher than those required to break through the blocking effect of the earth's field at mag lat 56° N (Bismarck), namely, 1·9 billion e-volts, enter the earth's atmosphere in considerable abundance, yet *there are practically no incoming electrons of lower energy than those that could just get through at that latitude**. These results are not at variance with those published in August 1938 by T. H. Johnson,† who in counter experiments extending to an altitude of 2 metres of water could not notice a difference between the counting rate at Minneapolis and at Churchill. Also, Carmichael and Dymond‡ reported in May 1938 that, in very consistent electroscope readings reaching altitudes of 12·7 mm of Hg taken within 5° of the north magnetic pole, they got not only the same shape of curve but also, within the limits of uncertainty of absolute measurements, about the same ionization which we observed at Bismarck and Saskatoon. Our own 1932 airplane observations in which we found no difference at 21,000 feet between Spokane and Cormorant Lake (p. 66) also bear on this point.

---

* Published first in Carnegie Institution Reports, *Year Book*, No. 36, p. 364 (10 December 1937). The italicized statement therein contained reads as follows: "We reach the conclusion that *practically the whole energy content of the latitude sensitive incoming rays is found in a band between* 3×10^9 *and* 17×10^9 *electron-volts*." See also *Science*, **87**, 427 (13 May 1938), and *Phys. Rev.* **53**, 855 (1938).

† *Phys. Rev.* **54**, 151 (1938).   ‡ *Nature*, **141**, 910 (2 May 1938).

Fig. 36 shows how one proceeds to obtain a quantitative comparison of the energies brought into the atmosphere by cosmic rays of varying energy. The area under curve D, Fig. 36, of course represents the background of ionization due to non-field-sensitive incoming radiation which is uniform the world over and upon which is superposed, to form curve C, all the additional field-sensitive rays (electrons) that can get through the blocking effect of the earth's field at San Antonio and produce ionization within the electroscope. The additional field-sensitive rays that at Omaha can get through the blocking effect of the earth's magnetic field are responsible for curve B. In a similar way, curve A is formed at Saskatoon. Then turning to Fig. 37, the three curves A, B, C of that figure represent, respectively, the differences between the curves C and D, B and C, and A and B of Fig. 36. The area under curve A of Fig. 37 corresponds to the total, or integrated, ionization produced per sec in each cc of the electroscope by all the incoming electrons passing through it of energies between 6·7 billion and 17 billion e-volts, the last two numbers representing, according to LeMaitre and Vallarta's calculations, the energy required by an electron to break through the blocking effect of the earth's magnetic field at San Antonio and at Madras, respectively, and to enter the atmosphere vertically. The mean energy, then, of the incoming electrons producing this ionization is, as aforesaid, about 10 billion e-volts.

Similarly, the area under curve B, Fig. 37, is the ionization due to the incoming band of electrons of energy between 2·9 billion (Omaha) and 6·7 billion e-volts, or a weighted mean of 4·5 billion e-volts. The area under curve C, Fig. 37, is the ionization due to the incoming band of electrons of energy between 1·4 billion (Saskatoon) and 2·9 billion e-volts, or a mean of 2·1 billion e-volts.

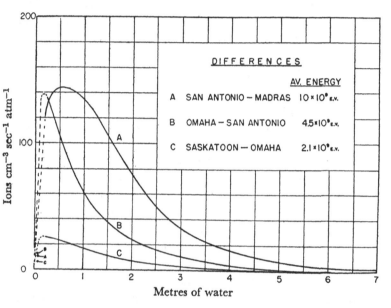

Fig. 37. The areas underneath these curves A, B, C are the same as the areas *between* the curves C and D, B and C, and A and B, respectively, of Fig. 36, and each area represents the energy brought in by electrons of the known mean energy 10 billion, 4·5 billion and 2·1 billion electron-volts, respectively. The points A, B, C on the *y*-axis are the computed values of ionization due to these electrons just outside the atmosphere.

In Fig. 38 the rectangular area 1 erected on the V (or e-volt × 10^9) axis is made proportional to the total or integrated ionization produced in the electroscope by all the electrons that enter between the latitude of Saskatoon

(60° N mag) and Omaha (51° N mag). This is the area underneath curve C, Fig. 37. Similarly, the rectangular area 2 is made proportional to the total ionization produced in the electroscope by all the electrons which get through the earth's field between Omaha and San Antonio. This is the area underneath curve B of Fig. 37. Again, the rectangular area 3 is the area underneath curve A, Fig. 37, while the rectangular area 4 is the area underneath curve D of Fig. 36, i.e. it represents the total ionization produced in the electroscope by all the rays of whatever nature, photons or electrons, which enter the atmosphere in the equatorial belt, i.e. at Madras.

The areas 1, 2 and 3 of Fig. 38 of course represent ionizations due to *incoming electrons* of energies between the limits shown on the V axis in the figure, but area 4, on the other hand, represents the total measured ionizing effect of all the rays that enter the equatorial belt, no matter what their nature may be. In so far as these rays are photons, we have no knowledge as to what energies are associated with them. We merely include them with the *electron rays* of energy above 17 billion e-volts (the part of area 4 underneath the dotted line) because they are found with them in the equatorial belt.

Having thus built up from the directly observed ionizations the rectangular areas 1, 2 and 3, we proceed without in any way changing these areas to readjust their shapes at the tops in the manner that is dictated by the single condition that there must be some *continuous* distribution of energies of the incoming electrons as their energies vary from 1·4 billion to 17 billion e-volts. This imposed condition leads to the final shapes of the areas 1, 2 and 3 as

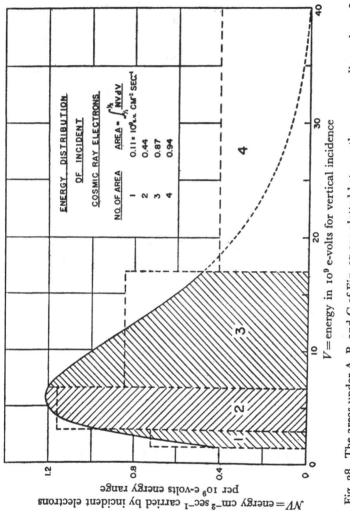

ENERGY DISTRIBUTION
OF INCIDENT
COSMIC RAY ELECTRONS

NO. OF AREA	AREA $= \int_{V_1}^{V_2} NV dV$
	$0.11 \times 10^8_{\text{v×}}$ CM^2 SEC^{-1}
1	0.11
2	0.44
3	0.87
4	0.94

$NV =$ energy cm^{-2} sec^{-1} carried by incident electrons per 10^9 e-volts energy range

$V =$ energy in 10^9 e-volts for vertical incidence

Fig. 38. The areas under A, B, and C of Fig. 37 are plotted between the corresponding values of individual electron energy, V, for vertical incidence. The ordinate thus gives the energy carried to the earth by electrons having energies lying between V and $V + dV$.

shown in Fig. 38, and it is notable how little flexibility in the shape of this curve up to the energy value 17 billion e-volts is left when the one condition of "no sudden breaks in the curve" is imposed. This curve then undoubtedly represents quite closely the actual distribution of incoming *electron* energies up to 17 billion e-volts. The extrapolation of this electron-energy curve beyond 17 billion e-volts (see dotted lines passing through area 4) has a reasonable chance of corresponding to reality, but obviously no certainty. As we have extrapolated it in Fig. 38, it takes care of about half of the integrated ionization which the electroscope actually experiences in the equatorial belt. The remainder of the observed ionization at Madras we have here represented by the remainder of the rectangle 4, which has been quite arbitrarily made to extend about as far to the right as the electron-energy curve extends before getting close to the *V* axis. This is more or less natural because of the rough interconvertibility of electrons and photons through nuclear impacts. Nevertheless, it is to be emphasized that rectangle 4 is inserted merely to have on the chart the representation *of the total ionization due to all the cosmic rays, no matter what their nature may be, that enter the equatorial belt*, and not to assert that the photon part of area 4 lies between the energy limits between which it is found in the figure. Where these hypothetical photons lie in the energy spectrum is of no particular importance for the present considerations.

There are certain definite conclusions that can be drawn from Fig. 38, as follows:

1. The first is that *the cosmic rays as they enter the atmosphere unquestionably have a definite banded structure*. This has

been pointed out repeatedly before, but never until now as the result of direct, indubitable energy measurements.

2. The second conclusion is that *the maximum of the cosmic-ray energy brought into the atmosphere by electrons in the northern hemisphere, where our measurements are made, lies at about 6 billion e-volts*, and that the energy distribution curve of the incoming electrons falls off rapidly on both sides of this point, dropping on the low-energy side, at 1·4 billion e-volts, for example, to less than a third of the maximum value, and on the high-energy side at say 20 billion e-volts, also to about a third of its maximum value.

3. The total cosmic-ray energy brought in by electrons of energy above 17 billion e-volts plus all that brought in by photons of all energies is about the same as the energy brought in by electrons alone of energies between 6·7 and 17 billion e-volts, and this energy is fully twice that brought in by all entering electrons of energies less than 6 billion e-volts. In other words, *the whole cosmic-ray energy comes in as a relatively sharply limited band.*

4. The smallness of the amount of energy brought in by photons, namely, only a fraction (probably not more than a half) of that represented by the area of 4, means definitely that *the entering electrons have not at all got into equilibrium with their secondaries before entering the atmosphere,* for in equilibrium Carlson and Oppenheimer have shown that "at any energy and thickness $t > 1$ ($t = 0·4$ m of water) there are always more γ-rays than electrons",* while in Fig. 38 the area assigned to photons is scarcely more than a sixth that assigned to electrons. This last conclusion does not rest solely upon the accuracy of the Carlson-Oppen-

* Carlson and Oppenheimer, *Phys. Rev.* **51**, 225 (1937).

heimer computations; for, as shown by the turnover points of the curves in Fig. 33, entering electrons even of a mean energy of 10 billion volts do actually get into equilibrium with their secondaries before they have penetrated even a twentieth of the way through the atmosphere, so that after it has become established that the entering particle rays are electrons* the smallness in the number of accompanying photons shows, from nothing more than a qualitative point of view, that *these rays cannot ever have come through an appreciable amount of matter in comparison with an atmosphere before entering the solar system.*

8. *The place of origin of the cosmic rays.*

The conclusion drawn in 4 above means that *the cosmic rays cannot have originated within the stars or in any portions of the universe in which matter is present in appreciable abundance.* This conclusion also appears to be indicated by the mere fact that the curve of Fig. 38 goes through a definite maximum at about 6 billion e-volts, unless the improbable assumption be made that the observed maximum is wholly due to the action on the incoming rays of the sun's magnetic field. For when an electron of given energy, say 10 billion e-volts, passes through matter, since the main mechanism of its absorption is first the formation of an impulse-radiation photon, then of an electron pair, then of two impulse-radiation photons, then of four electron pairs, etc., it follows that the energy corresponding to each value of V (Fig. 38) should remain a constant for all values of V lower than the original value of the incident electron energy. This permits of no such maximum as appears in Fig. 38,

* This was proved in *Phys. Rev.* **53**, 217 (1938).

so that if this maximum is inherent in the character of the rays as they enter the solar system, then no such process of degradation of energy through the "bremsstrahlung"-pair-formation process can have taken place. Further, if the original electrons had energies of many different values, some low, some intermediate, and some high, then the energy-distribution curve resulting from the passage of these electrons through a small amount of matter would be *one rising continuously with decreasing values of V*. The evidence drawn from the existence of this maximum appears, then, to be, in agreement with that drawn from the smallness of the photon component, that the incoming cosmic-ray electrons have not passed through an appreciable amount of matter on their way from their point of origin to the earth.

We have given attention to the question as to whether the influence of the sun's magnetic field could have been responsible for the appearance of the strong maximum at about 6 billion e-volts, as shown in Fig. 38, and have thought this unlikely from the consideration of the fact that if the blocking effect of the sun's field is not sufficient completely to prevent say 2 billion volt electrons from getting through to the earth (and we certainly find some of them getting through between 1·4 and 2·9 billion e-volts as Fig. 38 shows), then 5 and 6 billion volt electrons would probably get through to the earth, i.e. they could not be blocked off in appreciable amount by any sun's magnetic field which would let through even a small number of say 2 billion e-volts. But 5 and 6 billion volt electrons are both beyond the point of inflection that begins near the top of Fig. 38, and which then indicates a maximum, or *a banded structure of the incoming rays before they reach the sun's magnetic field at all.* This conclusion is at least not contradicted by the more rigorous computations of Dr Epstein,* who made a careful quantitative study of the effect of the sun's field on electrons coming into our solar system. On the other hand, the sun's magnetic field may well cut off largely from the earth electrons of original energies of say 2 billion e-volts, and it would probably block off entirely electrons starting toward the earth through that field if they have energies of 1 billion e-volts, or less.

9. *Speculations as to the mode of origin of the cosmic rays.*

A number of possible modes of origin of the cosmic rays have been suggested, and it will be appropriate to consider each of them in relation to the new data contained in Fig. 38.

1. It has been suggested that the energies of the cosmic rays are imparted by the fall of electrons through some sort of celestial electrostatic fields which thus impart the observed enormous energies of many billions of e-volts. This conception, difficult enough any way to reconcile with the uniformity of distribution of the incoming rays over the celestial dome, is also not easily reconciled with the fact that the energies of the incoming rays are limited to so narrow a range of energies as from 2 to some 20 billion e-volts. This form of origin would be expected to give a *continuously rising* curve with diminishing V in Fig. 38, since the electrons to be accelerated would normally be expected to be so distributed in the field as to take on all

* Epstein, *Phys. Rev.* **53**, 862 (1938). See also Janossy, *Zeit. f. Phys.* **104**, 430 (1937).

sorts of energies rather than energies in the near neighbour-hood of 6 to 12 billion e-volts. At any rate it is very difficult to get a banded structure with a maximum at about 6 billion e-volts out of such a conception.

2. Mr Hannes Alfvén has been trying to find the origin of the cosmic rays in the accelerating effect on electric charges of a pair of rotating double stars, each possessing a magnetic field.* Such a conception has as yet had no quantitative success, and it is difficult to reconcile with our conclusion that the incoming charged-particle rays contain no protons or other nuclei, but only electrons.† Also, it is difficult to reconcile with Fig. 38 because, since the charges to be accelerated would have to be distributed between the two stars and would thus be given widely different energies even when emerging from the same double star, not to mention the great diversities in double stars, so that a narrow band of energies such as we find, and above all an inflection point near 6×10^9 e-volts, would not be expected.

3. It was suggested years ago that the cosmic rays might be due to the partial or complete transformation, in accordance with the Einstein equation $mc^2 = E$, of the whole rest mass of the atom into cosmic radiation. If this transformation is assumed to be complete and the rays generated by this supposed transformation have suffered no degradation by passage through any matter at all—an extreme assumption—it is easy to compute the energies of the incoming rays from the fact that the mass of the atom of hydrogen is equivalent very closely to a billion

* Alfvén, *Zeit. f. Phys.* **107**, 579 (1937).
† Bowen, Millikan and Neher, *Phys. Rev.* **53**, 219 (1938).

e-volts. The energies released, then, by this sort of annihilation of the atoms of the most abundant elements found
in nebulae* (most abundant save for hydrogen and helium),
namely, boron, carbon, nitrogen, oxygen, aluminium and
silicon (none others are abundant enough to exert any
appreciable influence), would be, respectively, 11, 12, 14, 16,
27, and 28 billion e-volts. To correspond to the somewhat
extreme assumption of no degradation of energy whatever
in travelling from the place of origin to the earth such
complete annihilation of mass would have to occur, not in
the stars, where both temperature and density are relatively
high, but rather in the portions of space where matter is
not abundant. In any case, the momentum principle would
require that half of the energy shoot away from the point
of annihilation in one direction, and half in the opposite

* Bowen and Wise have just made by spectroscopic means a determination of the abundance of the elements in the nebulae and published the same in the *Bull. Lick Observatory*, **19**, 1 (1939). Their results
are shown in the accompanying table in which the figure given in the
abundance columns are the exponents required to indicate the number
of atoms in a given volume of space. Thus the number 11 following
Hydrogen means 10^{11} hydrogen atoms, while the number 9 opposite
Carbon means 10^9 carbon atoms etc.

Element	Abundance	Element	Abundance	Element	Abundance
H	11 −	Ne	8	K	6+
He	10	Na	≤7+	Ca	7−
Li	<8−	Mg	7+	Sc	<6+
Be	<8−	Al	<8−	Ti	<7−
B	<9	Si	≤9	V	<8
C	9	P	<8	Cr	<7
N	9−	S	8	Mn	<7
O	9	Cl	7+	Fe	7+
Fl	≤6	A	7		

direction. The energy of the cosmic rays shooting out in this way through the annihilation of the foregoing abundant common elements would then be a band of rays of energies lying between 5 and 14 billion e-volts—in other words, *in just the energy range in which the band of energies of the incoming rays are most abundantly found.* The whole curve of Fig. 38 would then be a reflection primarily of the abundance outside the stars of the different elements save for hydrogen and helium. But cosmic rays corresponding to the half-mass of the hydrogen atom, namely, half a billion e-volts, would in any case, according to Epstein's computations, be entirely cut out by the magnetic field of the sun, as would also all electron rays of lower energy. Also, the cosmic rays due to the annihilation of helium—2 billion e-volts—would be largely cut out in the same way. Indeed, above Bismarck (mag lat 56° N), where the blocking effect of the earth's field is very close to 2 billion e-volts, we have as yet found no increase in going north. At any rate, the distribution of energies of the cosmic rays shown in Fig. 38 would apparently not be irreconcilable with such an origin as that suggested above. If there is in fact the possibility of the complete transformation of the *mass* of a nucleus into cosmic radiation, i.e. into oppositely ejected electrons (or less frequently into two oppositely ejected photons), since only positive charges exist inside the nucleus, *the hitherto strange fact that the incoming electrons are certainly predominantly positives, quite possibly exclusively so, would perhaps be less surprising than it is at present.*

The question left not yet fully decided as to whether or not the maximum found in Fig. 38 is due to an inherent property of the cosmic rays, or is imposed by the action of the sun's magnetic field, should be definitely answerable through similar flights made in a sufficient number of intermediate latitudes.

It was as a beginning on this study that in late June and early July 1938 we made four successful flights in Bismarck, N.D. (mag. lat. 56° N) and three in Oklahoma City (mag. lat. 45° N) with the results shown in Fig. 39. The measurements obtained on different flights and different days at Oklahoma City were as consistent as those already shown in the tables on pp. 108, 109 corresponding to flights at Omaha and Saskatoon. But the flights at Bismarck on different days showed fluctuations greater than our observational uncertainties, as the group of curves in Fig. 40 reveal. This is presumably due to magnetic storms either on the earth or the sun. At the critical magnetic latitude of Bismarck the incoming electrons might well be peculiarly sensitive to magnetic changes in the earth, if 2 billion e-volt rays are present in abundance and trying to get through the earth's field. If on the other hand the sudden cut off near Bismarck is due to the sun's magnetic field, then solar magnetic storms would there cause fluctuations.

If now we pay no more attention to these fluctuations observed at Bismarck and, since in any case no more electrons are found entering at Saskatoon than at Bismarck, if we average the readings at Bismarck and Saskatoon, and compare them with all the data thus far taken at lower magnetic latitudes, we obtain the series of curves shown in Fig. 41. If then we treat these curves precisely as we treated those of Fig. 36, we obtain the apparent distribution of energies of entering electrons shown in Fig. 42. This figure seems to show, in addition to the incoming band of

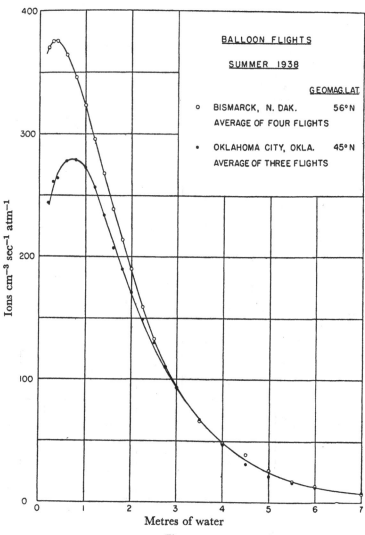

BALLOON FLIGHTS

SUMMER 1938

GEOMAG. LAT.

○   BISMARCK, N. DAK.   56° N
    AVERAGE OF FOUR FLIGHTS

•   OKLAHOMA CITY, OKLA.   45° N
    AVERAGE OF THREE FLIGHTS

Ions cm^{-3} sec^{-1} atm^{-1}

Metres of water

Fig. 39.

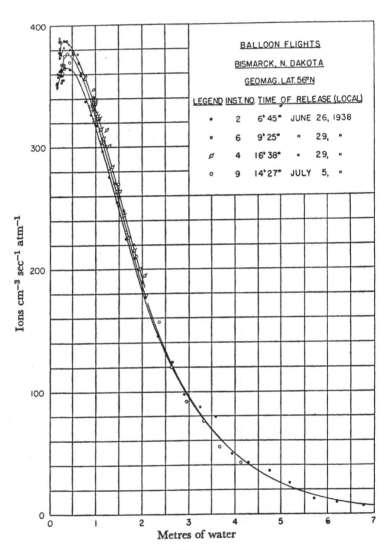

BALLOON FLIGHTS

BISMARCK, N. DAKOTA

GEOMAG. LAT. 56°N

LEGEND INST. NO. TIME OF RELEASE (LOCAL)

•	2	6ʰ 45ᵐ	JUNE 26, 1938
×	6	9ʰ 25ᵐ	" 29, "
⌀	4	16ʰ 38ᵐ	" 29, "
○	9	14ʰ 27ᵐ	JULY 5, "

Ions cm⁻³ sec⁻¹ atm⁻¹

Metres of water

Fig. 40

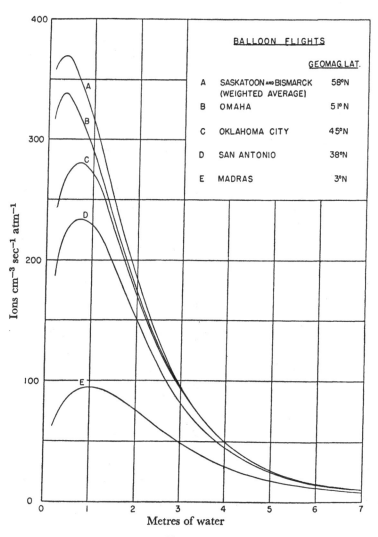

BALLOON FLIGHTS

GEOMAG. LAT.

A   SASKATOON ᴀɴᴅ BISMARCK   58°N
    (WEIGHTED AVERAGE)

B   OMAHA   51°N

C   OKLAHOMA CITY   45°N

D   SAN ANTONIO   38°N

E   MADRAS   3°N

Ions cm⁻³ sec⁻¹ atm⁻¹

Metres of water

Fig. 41.

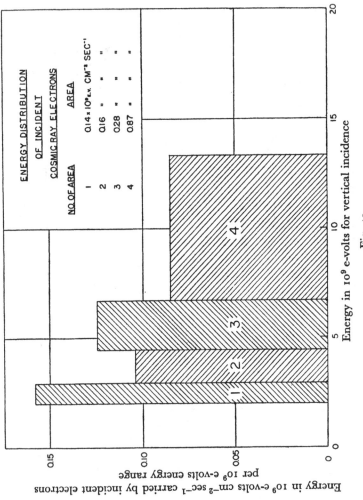

ENERGY DISTRIBUTION
OF INCIDENT
COSMIC RAY ELECTRONS

NO. OF AREA	AREA		
1	0.14 × 10⁹ ₑ.ᵥ. CM⁻² SEC⁻¹		
2	0.16	" " " "
3	0.28	" " " "
4	0.87	" " " "

Energy in 10^9 e-volts for vertical incidence

Energy in 10^9 e-volts cm⁻² sec⁻¹ carried by incident electrons per 10^9 e-volts energy range

Fig. 42.

electrons having a maximum at 6 billion e-volts, another band near 2 billion e-volts. This result is unchanged if, instead of using the mean of the readings at Bismarck and Saskatoon for obtaining the most northerly curve, we use for this curve merely all the readings taken at Bismarck alone.

This does not mean, however, that we as yet regard the existence of an entering band of electrons of energy near 2 billion e-volts as established. For by taking three flights at Omaha in the latter half of December 1938 we found the electronic energy coming in at that point some 9% larger in this winter season than we had previously found in mid-summer, the season in which all our preceding observations had been taken, and not all of it in the same year. These seasonal effects and the other fluctuations must be studied more fully before the point at issue can be definitely settled.

At any rate, however, these present experiments, taken in connection with Epstein's computations, seem at least to set an upper limit of 25 gausses for the strength of the sun's magnetic field; for such a field, according to Epstein's computation, should produce a cut-off of cosmic-ray electrons entering the earth's atmosphere at mag lat 58° N, about where our Bismarck experiments thus far seem to place such a cut-off.

It is quite probable, however, that the cut-off which we find at about the latitude of Bismarck is due, like the maximum at from 6 to 8 billion e-volts, to an intrinsic property of the incoming rays (the annihilation of helium atoms corresponding to just 2 billion e-volts) rather than to the sun's field; for according to the best averages we can now take of the thirteen most consistent Van Maanen measurements made at Mt Wilson by the Zeeman-effect method from 1914 to 1918* and the recent most careful measurements of Langer† by a modification of this technique (this brought out the existence of the field but gave it a very low value), the values of the strength of that field at the surface of the sun is probably closer to, say, 16 gausses than to 25 gausses. This would put the cut-off point due to the sun's magnetic field considerably north of Bismarck, in no case appreciably south of that point, and would lead us to seek another cause for the apparently observed cut-off near Bismarck. At any rate, in the present state of our knowledge it is legitimate speculation to regard the pronounced cosmic-ray band, possibly bands, between 1·9 and 20 billion e-volts shown in Figs. 38 and 42 as due to an inherent property of the cosmic rays, and possibly the apparent cut-off at Bismarck as well, and to ask what kind of natural process can yield such stupendous energies. The annihilation process mentioned above seems to be the most plausible one now in sight. If it is correct, it is obviously of fundamental importance.

I hope I have been able in these lectures to show that there is both interest and profit in the study of the cosmic rays.

* See Review by Frederick H. Seares, *The Observatory*, No. 556 (20 September 1920). See especially p. 318.

† Carnegie Institution of Washington, *Year Book*, No. 35, p. 173 (1936).

INDEX

Absorption coefficient, 85, 102, 103
Absorption law, 98, 102; of electrons, 54
Alfvén, 121
Alpha-particle track, Fig. 11
Altitude ionization curve, 87
Alvarez, 79
Amsterdam, 72
Anderson, Carl D., 28, 30, 31, 38, 41, 42, 44, 45, 46, 48, 51, 96, 100, 80, 85, 95, 96, Figs. 3 and 9
Ångström, 15
Astronomy, what is it good for, 3
Auger, 39, 94
Authoritarianism, 5, 7

Balboa, 58
Ballot government, 7
Barnothy, 28
Bartol Foundation, 79
Batavia, 60, 72
Beta-ray tracks, Fig. 2
Bethe-Heitler theory, 35, 43, 45, 46, 47, 48, 49, 54, 84, 86, 92, 94, 98, 100, 101, 102, 103
Bhabha, 92
Bismarck, 111, 123, 124, 129
Blackett, 31, 32, 41, 42, 45, 96
Blocking effect of earth's field, 72, 112
Blocking effect of sun's field, 119
Bolivia, 59
Bothe, 61
Bowen, 18, 46, 63, 64, 83, 86, 121, 122
Braddock, 38
Bremsstrahlung, 48, 54, 84
British Association, 58
Brode, R. T., 41, 53, 96
Brookings Institution, 5
Bruins, 69
Bursts, 100

Cameron, 18, 19, 40, 41, 57, 85
Cape of Good Hope, 72
Carlson and Oppenheimer, 92, 99, 117

Carmichael, 100, 111
Carnegie, 60
Carnegie Corporation, 63
Carnegie Institution, Fig. 20, 78, 86, 111, 130
Celestial mechanics, 3
Chadwick, 32
Chao, 83
Churchill, 61, 67, 111
Cicero, 22
Clay, 39, 60, 63, 69, 72, 75, 80
Cloud chamber, Fig. 3, 27, 29, 41, 53, 70, 71, 85, 94
Coefficient of absorption, 44
Colon, 30
Communications, Art of, 11
Comparison of flights at Saskatoon, 108
Comparison of ionization curves, 127
Compton, A. H., 26, 31, 63, 64, 67, 69, 75, 79
Core of knowledge, 20
Corlin, 61
Cormorant Lake, 63, 66, 67, 69, 111
Corson, 53
Cosmic rays, banded structure of, 116; discovery of, 6, 8, 19, 28; edge of plateau, 64; energy distribution of, 105, 128; energy distribution of c.r. electrons, 114, 115, 116; energies, 29, 56, Fig. 3; equatorial dip of, 64, 75; intensity change, 77; maximum of, 117; metre, Figs. 24 to 27; mode of origin of, 120; number entering per sq. cm., 106; number of per sq. cm., Fig. 7; penetrating power of, 39, 43; place of origin of, 118; range of, 39; range properties of, 70; record, Fig. 31; showers, 35; small photon component of, 119; total energy of, 107; what they are good for, 1, 2, 8

Printed in the United States
By Bookmasters